Irrigation Management

DPH Management Series

Irrigation Management

J M DEWAN • K N SUDARSHAN

DISCOVERY PUBLISHING HOUSE
NEW DELHI-110002

Discovery Publishing House
4831/24, Ansari Road, Darya Ganj
New Delhi - 110 002 (INDIA)

Irrigation Management

Reprint : 2012
ISBN-81-7141-369-2

PRINTED IN INDIA

Published by Discovery Publishing House, New Delhi and Lasertypeset at Authorsgroup and Printed at Dynamic Printers

Contents

Contents

Preface

The management world is in transition. The causes of this transition are many, but the major one is the vast changes in knowledge and in the information that flows in and out of organizations. This changing information disrupts traditions, established processes, well-known procedures, and routine ways of doing things. New principles, concepts, techniques, ideas, expressions, processes, and procedures are emerging, moving us to a new plateau of professional practice. Trying to capture this changing knowledge and information is like trying to capture the atmosphere. How can you do it when the atmosphere is continually shifting and when you need the atmosphere to do the capturing? The best we can do is find a peak from which we can at least get a perspective on management as a whole, decide on the work and responsibilities of management, and gather in whatever practical management information we can. A team of experts in this series represent some of the best contemporary thinking and information available. They represent many major successful corporations, active consulting agencies, and well-known educational institutions, and all are experts on what is happening with the flow of knowledge and information in the management world. This is a lofty pinnacle from which to survey the management world.

Managers and supervisors clamor for current information and guidelines to help solve formidable problems in their work world—problems that range from "how to do it" to "how to resolve conflict when doing it." Many problems are generated from miscommunication and incompetence. As the management practice proceeds from the complex to the supercomplex, problem solving becomes a large-scale challenge requiring new knowledge and skills. Managers and supervisors cannot wait for research breakthroughs with real-world answers to solve these dilemmas. They must tackle them here and now with the useful information and proven practices immediately available. Whether making a decision, solving a problem setting up a procedure, designing a process, or resolving a behaviour conflict, a manager must rely heavily on information. To a great extent, management practitioners are information workers; that is, they generate, distribute, store, retrieve, and consume information. Competence in finding and using the right information at the needed time determines to a considerable extent competence in the management function, activity, or responsibility. The *DPH Management Series* attempts to fill this need for usable information in spite of the changing nature of its subject.

The *DPH Management Series* not a book to be read and later discarded. It is a reference book, a tool to be used by managerial personnel in the day-to-day work of an organization. Like a tool, it should never be more than a reach away when a new situation emerges that demands its use. This series aim to achieve a first-and practical and proven knowledge and information as a self-development opportunity for those who are moving into or upward in management. A complete spectrum of management subjects is immediately available for orientation, study, analysis, assimilation, and problem solving. Within one set of covers is the view of management as a totality. The management field is loaded with ideas that the organization of this handbook series unique logic. It follows both levels and areas of responsibilities of an organization.

The work of this handbook series is the collaborative effort of many outstanding people in the management field. The motivation for this work varied from individual to individual, but the central motivation that united us all was the excitement of capturing the management state-of-the-art and sharing it with colleagues in the dynamic profession of management.

This series should be of great help to managerial practitioners at any organizational level who are responsible for a function, department, or set of responsibilities. The handbook series will also give these practitioners insights into management roles and approaches in other areas as well. The subject matter encompasses top, middle, and lower management. Special emphasis was placed on managing people, time, space, budgets, and resources to give the handbook extra utility for middle and lower management. Students of management in university or educational institutions will find the series an invaluable resource for adding "real world" practices to their academic and theoretical foundations. MBA students will gain an invaluable overview of the total organization to complement their MBA degree. Administrators and public managers can become acquainted with practices employed by managers and supervisors in private organizations. These practices are not always directly applicable in public sector bodies, but with thought and modifications, these private practices can adapt to public organizations. Public and university librarians will find the handbook an indispensable reference for the multitude of questions on many topics from the general public, special groups, associations, and students.

Editors

1

System Analysis and Irrigation Management

What system means

The word "System" is composed of two Greek words viz. "Syn" and "histana" meaning 'together' and 'to set' respectively. Thus, the literal meaning of "System" means 'to set together'. Hence, it may be more appropriate to say that a system is an interdependent and inter-related set of components. In other words, system is a group of inter-related components acting for a common purpose.

System defined

According to the dictionary meaning, System is "assemblage of objects arranged after some distinct method usually logical or scientific, whole scheme of created things regarded as forming one complete whole".

Stafford Beer defines the system as "Anything that consists of parts connected together".

Mac-Farlane, A.G. J. defines system "An ordered arrangement of physical or abstract objects".

Ackoff, R.L. defined system as "Any entity

conceptual or physical, which consists of independent or physical and that the system consists of interdependent parts".

Drenick, R.F. defines system as "A device which accepts one or more inputs and generates from them one or more outputs". This definition links up inputs and outputs and is commonly used.

James C.I. defines system as 'Any structure, device, scheme or procedure, real or abstract, that inter-relates in a given time reference, an input, cause or stimulus, of matter, energy or information, and an output, effect or response, of information, energy or matter". This definition provides emphasis on the function of the system that interrelates, in some time reference, an input and an output.

System analysis

The concept of system analysis is to focus on the system under consideration as a whole, instead of each of its individual components, in order to improve the planning, design or operation of the entire system.

Biswas, A.K. defines system analysis as an explicit analytical study that assists any decision makeer to select a preferred course of action by identifying and examining the possible consequences of several feasible alternatives. It is a logical and systematic approach wherein assumptions, objectives and criteria are enumerated. The technique can help the decision maker to arrive at better decision through the developed framework for analysing the problem rather than *ad hoc*, piecemeal approach as often attempted.

The system analysis of irrigation management may be

viewed from this angle as an integrated approach to identify and then rectify the existing lacunae of the system to attain maximum water use efficiency and overall increase in agriculture production.

The details of system analysis

It is the art and science of selecting from a large number of feasible alternatives. The set of actions which best accomplish the overall objectives.

Three phases of System Analysis

1. Problem formulation (Art)
2. Problem of Analysis (Science)
3. Evaluation of Analysis (Art & Science)

Steps in System analysis

1. Identify goals (Direction of change)
2. Identify objects (Quantitative measure of change or index of perfromance)
3. Formulative alternatives
4. Formulate models (Conceptuation of a system while retaining essencial characteristics of that system for a purpose)
5. Select best solution

Advantages of System Analysis

1. Systematic approach
2. *Focus on the overall problem*
3. Interdisciplinary approach
4. Promotes/motivates(Reactivity)

5. Solution sensitivity to uncertainties.

Formulation of the optimization problem

Max (Min), $Z (X_1 X_2 ... X_n)$

Subject to (S,t)

$I_1(X_1, X_2 ... X_n) \leq G_1$

$I_2(X_1, X_2 .... X_n) \leq G_2$

$_3(X_1, X_2 ... X_n) \leq G_3$

$I_m(X_1, X_2 ... X_n) \leq G_m$

Here, $X_1, X_2 ... X_n$ are=Decision variables

$Z/Z(X_1, X_2 ... X_n)$ are= Objective Functions

$I_1(X_1, X_2 ... X_n)$ are = Constraints

Basic method of optimization

Identify major sub-sets of the decision variables which cannot contain the optimizing set.

Example

Goal—Optimize the value of your time

Define

X_1 = Hours of work

X_2 = Hours of sleep

X_3 = Hours of eating

X_4 = Hours of recreation

Rank your preference

$X_4 \Rightarrow$4-Most value

$X_3 \Rightarrow$3-Good value

$X_2 \Rightarrow 2$ Moderate value

$X_1 \Rightarrow 1$-Least value

Max. $Z = 2\,X_1 + IX_2 + 3X_3 + 4X_4$

S.t. $= X_1 + X_2 + X_3 + X_4 = 24$

Solution if $X_1 = 0, X_2 = 0, X_3 = 0, X_4 = 24 ... Z = 4(24) = 96$

Max $Z = 2X_1 + 1X_2 + 3X_3 + 4X_4$

S.t. $= X_1 + X_2 + X_3 + X_4 = 24$

$X_1 \geq 6$

$X_2 \geq 7$

$X_3 \geq 2$

Solution — $X_1 = 6, X_2 = 7, X_3 = 2, X_4 = 9$

$\therefore Z = 61$

Max $Z = X_1 + X_2 + X_3 + X_4$

S.t $= X_1 + X_2 + X_3 + X_4 = 24$

Solution $Z = 24$ for all

Feasible Combinations of X

What went wrong?

Value is no — adequate

Measured by ranking

And it is a function of the amount of decision variable X.

Optimization methods

1. Non-linear Optimization & Linear Programming simple (i.e. based) algebric eq, $y = m$, $X = 0$) Optimization with Calculus

It is based on derivatives, hence derivate must exist, function must be continued/continuous and differentiable

2. Numerical search method
3. Deterministic Optimization
4. Stochastic Optimization
5. Dynamic Programming

These methods are not frequently used because of

(a) Lack of adequate training in the effective use of these models

(b) More than just making the model run

Thus,

Single variable linear programming

e.g. I — $Z = (x - 5)^2$

Variable 2—(Non-Linear). $= 2(x-5) = 0$

$\therefore x = 5$

$\frac{D^2z}{dx^2} = 2 \Rightarrow$ Minimum

Simulation models

1. Predicts impacts
2. Identify data needs
3. Store data(Data Bank) $\frac{dz}{dx}$
4. Identify physical and institutional relation—ships needing research
5. Identify system objectives

6. Training mechanism
7. Easy to assess whether output is reasonable

Types of mathematical methods

1. Simulation
2. Optimization—Descriptive and Prescriptive
3. Steady flow
4. Unsteady flow
5. One-Dimensional
6. Two-Dimensional
7. Three-Dimensional
8. Quantity-quality
9. Daily
10. Monthly
11. Yearly

Optimization models

1. Identify best actions
2. Identify system objectives
3. Identify preferences between objectives
4. Identify impacts of system constraints
5. Hard to assess whether output is reasonable
6. Objectives
7. Optimization method.

Simulation model

Simulation models use mathematical equations to express

the physical interactions of a river-reservior system or any other live systems. These models help to evaluate the impact of a given backfeed or input in respect of the system (e.g.) failure of a dam or building etc. by hypothetical happening of any incidence or event. This is done by preparing a questionnaire about the happening of the event under certain set of conditions. It has been experienced that such simulation models can be used easily and conveniently to draw conclusions to the nearest of actuality of the conditions under which actual happening is likely to occur for a given system. The technical feasibility, reliability of the output results can be judged. The simulation models interpretations are judged for their reliability by comparing them with optimization models. These models have been found to be quite reasonable. By these and interpreted accordingly. The type of various technical data or all needed informations of the system for making simulation models are framed and feeded to computers. Simulation models have been found to be quite effective to evaluate physical, institutional and other interactions which may be required for advanced research studies. These models are also effective tools of training to in-service or freshers of any profession.

For the purpose of irrigation system it is possible to ascertain reservoir release conditions based on probable storage, inflow and criteria of reservoir operation simply by plotting the hypothetical model release quantum and desired release levels. The quality planning design, water location size etc., simulation models, extent of seepage, leakage, and efficiencies, river basin models can be made. The river conditions below and between reservoir, diversion and confluence points the river segments connecting the nodes, mass balance, energy relationships and decision making etc. are the other benefits which could be drawn from such models.

The model outputs for different alternatives are worked out and most effective and economical can be chosen for a given system.

Optimization model

The optimization models are essentially simulation models with a built in mechanism to generate and evaluate alternative solutions in a systematic fashion so as to identify the most technically reasonable, cost effective and workable alternative methodology to accomplish the job. In a complex system where many informations and alternatives are required, the number of models are to be generated and choose the one which suits to policy planners and technocrats. The choice of a model is also done by trial and error method. Conceptually optimization models help to eliminate, delete and minimize the difficulty in choosing a desirous alternative solution.

The optimization modes are expressed in mathematical forms or relationships and interpreted by the experienced professionals. In case of a complex system it is also preferred to optimize the single objective of that system viz., optimizing the hydropower generation of a multipurpose irrigation project or likewise other objectives of that system e.g. flood control, municipal or industrial use or recreation, etc. However, while adopting this technique the care should be taken that the particular object analysis done through single optimization model does not decrease the importance of other objects of the system to keep its framework of multipurpose project for this enough experience, skill and training is needed. This constraint of optimization model is no faced in simulation models and hence many prefer for later models as well. A good planner works out all kinds of models for taking best decision from various alternatives cooked our from different models which meets the requirements to a reasonable level or extent. This is termed as the articulation of the objectives of the projects and in fact is more important than the results worked out from different models.

The optimization models are highly mathematical as compared to simulation models but the former provides a mathematical algorithm to generate and discriminate among various alternatives depending upon the techniques e.g. linear, non-linear or dynamic programming etc. The effective use of an optimization model needs the basic know-how of the physical system's nature as well as solution characteristics which are to be simulated and optimized.

The trial and error technique for working out suitable alternative is also done in optimization models like the simulation models. The maximization or minimization of priorities of the given object in a given project is done while keeping other objectives at specified threshold values/limits. The process is repeated till the most workable alternative is obtained keeping in view the various constraints of the project to meet the goals.

Thus, the simulation models help the manager to better understand the system whereas the optimization model pinpoints the better articulation of the objectives to successfully and correctly accomplish the job. The backfeed of simulation models are the purposes/objectives of a given system while that of optimization is the full description. These tools of analysis go simultaneously like a hammer and saw of the carpenter i.e. both have their own importance and need to accomplish job. However, it is again emphasised that use of mathematical models need spatial skill, training, practice and intelligence to use them with perfect ability.

Hybrid models

The complex projects involve the mixed methodology of two or more models to set the best alternatives. Such models when developed can be termed as hybrid models.

The design, analysis and management of water resources systems are highly complex due to the unique nature of the resource water. The large scale of the system, the non-linearities of the relationships exhibiting interactions within system and with the environmental factors, the stochastic nature of inputs and outputs, the different constraints influencing the resource use, their analysis and the time staggering or leg in a few aspects of the behaviour of system makes it a difficult model to interpret perfectly. Moreover, these systems are example of long term public investments and are irreversible in nature. Analysis of such systems, therefore help in making correct decision for their effective operation. The analysis of these public systems, intended by the governments to serve national objectives of economic efficiency, income redistribution or proper recycling of money among masses, regional development, environmental stability and general social well being, the various problems encountered by the systems are to be studied and interpreted carefully to explain the evaluation of these multiple objectives systems. However, views may differ about these objectives and criteria for irrigation and irrigation management. The difference of opinion may be on individual, group, discipline, professionalism, departmental or state levels regarding the consideration of the one or several objectives. The criteria are considered so as to evaluate who is being benefited and where objectives and criteria are located on long casual chains.

Irrigation management has been defined as a process of manipulation of water resource to utilise it in the higher production of food and fibre. In actual sense of the word it also includes the management of water in rainfed areas too besides that provided for perennial irrigated areas by man made irrigation systems of different type and size. Irrigation management neither includes water resources, dams or reservoirs to store water nor the water conveyance system, nor codes, laws or institutions to allocate water, nor farmers organisations, nor soils or cropping systems or drainage systems but it is the way by which certain skills and organisational approaches are used to manipulate physical, biological, chemical and social resources in order to bring the water to farms and crops for deriving more food and fibre production.

It has been suggested that for a practical framework five local objectives may be considered:

(1) Water Productivity;

(2) Equity in distribution;

(3) Stability for long term both environmental and maintenance of works;

(4) Carrying capacity—comprising livelihood intensity reflecting the size of population at different levels; and

(5) Well beings—including health, amenities, nutrition and psychic factors.

Analytic optimisation techniques such as linear

programming require to formulate a problem in particular formate. Measures to optimise achievement of the said objectives water, choice, of cropping system and crop zoning as per soil and irrigability patterns in the area, the frequency with which different zones receive irrigation, the staggering of cultivation and spatial as well as temporal spread of cultivation rights. This would result in a few serious conflicts between the objectives of a system or systems. The main conflict visualized is between productivity and equity under circumstances where high transmission losses decrease the productivity of any system, the distribution of water to tail enders. However, this may be disturbed by the highly productive use of ground water which is being recharged by seepage losses from the system. The complementaries of such conditions are more beneficial especially in respect of redistribution of water from head to tail which can be instantaneously more productive, more equitable, decrease the waterlogged and water-stagnant conditions, support more livelihoods and increase opportunities of social well beings.

The optimisation process, however, involves many limitations like who is more or less benefited from the practices introduced now and in long term. Therefore, in view of these conditions the approach of practical political economy is suggested which seeks realistic approach to analyze how all can gain and how losers can be reconciled to their losses. These may be achieved in three ways namely the professional training and incentives of irrigation managers; the search for ways in which tail reach farmers can gain while receiving less water and the purchase and distributions of land to landless and very small farmers at the time of the commencement of irrigation in newly created projects.

Ministry of Water Resources, Govt. of India had set up a working group in 1986 to prepare guidelines for project reports on irrigation systems. The committee has given recommenda—tions to provide informations as given below:

(1) Submergence, rehabilitation, compensation etc.

(2) Design assumptions criteria, data and investigations carried out.

(3) Geological investigations of the foundations of the various structures.

(4) Counter plans of the commanded area.

(5) Ground water levels and possibilities of water logging, arrangements for drainage.

(6) Soil surveys and suitability of soil for irrigation.

(7) Conjunctive use of ground water and surface water for optimum use of available waters.

(8) Conveyance losses in canals.

(9) Agreements on interstate aspects on utilization of waters, submergence etc.

Bottral pointed out other limitations of design which are:

(1) Planned limitations.

(2) Deficiencies in water courses layout.

(3) Insufficiency provision for drainage.

(4) Absence or shortage of water measuring devices.

System analysis has been evolved from interdisciplinary team which synthesize the techniques and viewpoints from their several disciplines. It is worked out through analysis of existing or proposed programmes or systems quantitatively based on their performance and later evaluating it as a whole for deriving suggestions and conclusions for judging the adaptability under various set of conditions.

This may involve three steps namely:

(1) System orientation i.e. analysing system as a whole considering all influencing components or factors avoiding the single component approach being practised in the past.

(2) Interdisciplinary approach to system considering variables equally important.

(3) Working out formal mathematical models through association of natural or basic sciences.

The system models help to bridge communication gaps and coordinate various interactions and field informations. The systems being described and discussed in this book are primarily water and related land resource. These are systems in which capital investment, its regulation and management factors are involved. These further are subjected to public intervention on account of the involved imperfections likely to come up due to several human factors and those concerned to market system and economies of scale.

System analysis is not just a mathematical model or technique which can replace the judgement and professional experience of managers and political decision makers. The technique of system analysis involves a rational basis for performing an action programme on any selected methodology out of many available alternatives according to its likely suitability and viability for its higher efficiency. The planning and operation of water resource system can be effectively improved, if the managers as well as the analyst interact favourably to understand viewpoint of each other regarding the lacunae and constraints of the approaches choosen for system analysis. The use of system analysis in water resources planning has started playing a major role and is likely to expand in future.

System management

Management of any entrepreneur or system is a concept an mental attitude towards efficient benefits or achievements. It exercises a basic systematic technique or approach for discovering or identifying or diagnosing or pinpointing the problems or lacunae which increases the efficiency of a system. It results in higher benefits through sealing of leakages of any kind occurring in system due to the reasons identified and by exemplifying the best use of human knowledge and efficiency as well as material energy. A general diagram explaining the systematic analysis of any system or enterprise of any kind may be drawn as below:

The operation of laws of psychology
Sociology

Environmental of high moral

Physical, Chemical and Climatic

Biological laws, statutes, time limit
OBJECTIVES
PLANS
STANDARDS
METHODS
SCHEDULES
AND
CONTROLS
OF AN
ENTERPRISE/SYSTEM
all within laws of each situation.

constitute the laws of the situation

Tradition, Spiritual and human values

Thus for any system analysis and for achieving efficient, and economical benefits from it such diagram should be evolved where in the INNER-RECTANGLE should indicate constructions or limitations which are not to be violated or ignored or breached natural laws, money i.e. economical aspects, and time limitations for achievements and statutes. Around the LARGER RECTANGLE should be shown or considered those factors of parameters or conditions that is applicable to system or enterprise and the peoples associated with it and those being served or involved in drawing benefits such as farmers in irrigation system.

Having identified the limitation within which management or manager or team of workers or team of experts etc. must work, then it is easy to make out an objectives, plans, standards, methods, schedules, subsystems, involved, interdisciplinary approach for various disciplines involved for complete control of enterprise or hold on the system. When this approach is fully understood, agreed and accepted by the entire team working for management, it would become way of life for any organisation whether commercial, technical, social or of any type. It thus brings about the cohesion of thoughts and understanding to work in the similar fashion to achieve goals determined. This team then certainly can state that "we analyse a situation or system to a point where the system facing a problem or situation of any kind to be improved comes to order as per expectations.

The chart suggested is a model of system analysis i.e. it may be considered as a guideline to understand the system. However, this is to be made and interpreted separately for the different systems or enterprises as per needs. Thus this would work as a guiding management tool to be used with considerable skill and ability. The laws of the situation if taken for granted such as in the case of irrigation system the effective use and equitable distribution of water is disobeyed the people loose their faith in management and starts playing with the system which recks the system and it looses its efficiency, benefits and thus becomes utterly failure and liability on the society. Besides other serious consequences of the society a whole has to bear such as waterloging, salinity and/or alkalinity, environmental hazards, decrease in production, poverty and ultimately slavery etc.

The various factors given in chart should be elaborately defined for a particular system. Thus objectives plans standards, methods schedules and controls

The system analysis of any enterprise much depends upon its supervisor or incharge who should have art to achieve the benefits.

The few of the characteristics of a good manager or captain or leader of the enterprise or team should be following:

(1) Divide task for distribution amongst team workers.

(2) Develop team work through mutual understanding.

(3) Collect ideas regarding best accomplishment of the job.

(4) Make suggestions for improvement.

(5) Ask questions from fellow team workers.

(6) Listen with patience the thoughts/ideas of his team workers.

(7) Encourage initiations for better work through rewards, certificates etc.

(8) Recognise the good work of the entire team and of individuals.

(9) Poor work should be constructively criticized to avoid its repetition.

(10) Prevent wasteful expenditures.

The importance of these guidelines is much more during these days because due to highly specialized techniques and experts for that are to be coordinated. Any mistake on the part of coordinator or leader or incharge would ruin the project work.

Interdisciplinary dependability

Irrigation management is a interdisciplinary subject wherein engineers, hydrologists, hydro-geologists, meteorologists, crop scientists and growers, soil scientists, economists and social scientists are involved at one or the other stage of water resources development and utilization in agriculture. An integrated approach is, therefore, needed if the available water resources of the country are to be used efficiently in crop production. For instance that average yield of wheat in Punjab has reached to a level of 22 quintals per hectare with about 87 per cent of the wheat are irrigated. In Maharashtra and Madhya Pradesh where irrigated wheat is much less, the average yields are low. However, the availability of water for irrigation is not sole factor for increased production. This is obvious from the fact that Haryana and U.P. having created high percentage of irrigation have not registered increase in production as compared to Punjab. This point reflects to the fact that for optimal water use efficiency a complete crop production technology is needed in which water is one of the important components.

The advent of irrigation results in the changes of technology, depending upon the availability of water e.g. change in cropping patterns selections of crop species and allied aspects. Thus in irrigated areas the cultivation of paddy increases many fold within few years due to abundance of water required for this crop. Thus crop such as maize etc. which are susceptible to water stagnation and water logging the cultivation of those is either reduced or abandoned in Kharif. In West Bengal in winter both rice and wheat can be grown but the water requirement of former being high has to be met according to season hence its area in winter has to be limited if at all this crop is grown.

Thus it is obvious that besides the creation of irrigation potential what is equally important is the management of water. Sine water is one of the constraints during rabi cultivation, a farmer has to decide on the cropping pattern based on the water needs of the various crops as well as economics of the crops.

There is about 70-75 per cent rain fed cultivable land in India, in about 40 per cent the crops do not experience water deficit during their growth. However, there are regions which often suffer because of the excess of water resulting in problems of drainage and water logging, Rice is the predominant crop of these regions. Fluctuation in this commodity would affect the national grain policy as it accounts to about 37% of the grain production, water management in these areas provides a key to improvement of yields of rice and ultimately the national production.

Out of the 75% rainfall area about 35% area though receives 250 to 1000 mm but it is uncertain and erratic, resulting in periods of water sufficiency and deficiency.

These areas need cumulative efforts of water, harvesting, moisture, conservation and its efficient use till next monsoons.

On farm water management may be successful and efficient only when the clear understanding of the soil characteristics, the hydrophysical properties such as water retention and transmission fertility are well understood.

The recent advance of water use technology suggest that there is a differential water use efficiency in different crop plants. With the development of crop canopy and change in the season, the crop water requirement change. In this process there is a possibility that in a particular cropping pattern chosen by the farmer, all crops may require water at one time. This can reduce water use efficiency if water becomes a limiting factor. Therefore, the evolution of cropping pattern has to take into account the availability of water at different stages of plant growth on one side and the requirement of water by different crops on the other. This can help in predicting a cropping pattern which will be economical to a farming system. The various interdiscplinary disciplines which influence water and land system resource have been discussed in further chapters considering their direct or indirect influence on irrigation management.

Showing state-wise potential created, utilisation and Lag in utilisation

Sl. No.	State/union territories	Lag at the end of 6/79			Percentage of Lag	Remarks
		Potential (utilisation Lag by 6.78 upto 6/79)				
1	2	3	4	5	6	7
1.	Andhra Pradesh	2783	2756	27	0.97	-

2. Assam	61	41	20	32.78
3. Bihar	2302	1640	662	28.76
4. Gujrat	957	536	421	43.99
5. Haryana	1710	1602	108	6.31
6. Himachal Pradesh	-	-	-	-
7. Jammu & Kashmir	100	95	5	50
8. Karnataka	1008	1010	-	-
9. Kerala	432	439	-	-
10. Madhya Pradesh	1256	1026	230	18.31
11. Maharashtra	1128	667	456	40.60
12. Manipur	-	-	-	-
13. Meghalaya	-	-	-	-
14. Nagaland	-	-	-	-
15. Orissa	1326	1392	-	-
16. Punjab	2258	2274	- -	
17. Rajasthan	1376	1207	169	12.22
18. Sikkim	-	-	-	-
19. Tamil Nadu	1178	1165	12	0.12
20 Tripura	-	-	-	-
21. Uttar Pradesh	5472	4677	795	14.71
22. East Bengal	1420	1439	-	-
Total (States)	24757	21967	2905	11.73
(Union Territories)	10	10	-	-
(All India)	24767	219777	2905	11.73

Showing creation of irrigation potential and its utilisation
(Area in ha)

Year	*Potential*	*Utilisation*			*% of utilisation*	
		Kharif	*Rabi*	*Total*	*Kharif*	*Rabi*
1975-76	21000	-	87485	8785	-	41.83

1976-77	22000	-	13175	13175	-	59.88
1977-78	46000	-	10965	10965	-	23.84
1978-79	65000	-	14009	14009	-	21.55
1979-80	83000	4176	27061	31237	5.03	32.60
1980-81	97000	9722	47330	57052	10.02	48.79
1981-82	147000	1338	46716	48054	0.91	31.77
1982-83	182000	5843	61562	78405	3.21	33.83

Irrigation potential in M.H.A. (cummulative levels

Item	Ultimate	1950-51	1960-61	1968-69	1973-74	1977-78	1979-80
Major Medium Irrigation	58	9.70	14.30	18.10	20.70	24.77	27.02
Minor Irrigation							
(a) Surface	15	6.40	6.45	6.50	7.00	7.50	8.00
(b) Ground-water	40	6.50	8.30	12.50	16.50	19.80	22.00
Total Minor irrigation	55	12.90	14.75	19.00	23.50	27.30	30.00
Total irrigation	113	21.60	29.05	37.10	44.20	52.07	57.02

Showing food production-1951 to 1979 (million tons)

Year	Production	Year	Production
1951	55.0	1966	72.3
1952	55.6	1967	74.2
1953	61.8	1968	95.1
1954	72.3	1969	94.0
1955	70.7	1970	99.5
1956	69.3	1971	108.4
1957	72.5	1972	105.2

1958	66.6	1973	97.0
1959	78.8	1974	104.7
1960	77.1	1975	99.8
1961	82.3	1976	121.0
1962	82.4	1977	111.2
1963	80.3	1978	126.4
1964	80.7	1979	131.4
1965	89.4	1980	

2

Groundwater Markets and Sustainable Water Resource Management

Private exploitation of groundwater involves powerful and extensive externality effects. Some of these directly hit the poor; some affect all owners of land in certain regions by reducing or enhancing the long-term productivity of land. This chapter attempts and analysis of three situations in which externality effects of ,groundwater exploitation have become important in the Indian context. We shall also attempts to identify areas where significant income distributive and environmental gains can be achieved by modifying the institutional arrangement, or by a judicious mix of suitable tax-subsidy-type public interventions, or through more effective macro-level resources management.

Well interference and the mining of groundwater

The relationship between the water output of a well and its area of influence is governed by established hydraulic principles. The yield of a well is determined by hydro-geological parameters such as the ease and speed with which water can move through the aquifer, the drawdown available in the well, etc. In the traditional water extraction technology regime, water output per well was low because of stringent constraints on the quantity of water that could be lifted per unit of time with human and animal power; as a result, well interference was non-existent and only a fraction of the utilizable groundwater potential was used. Modern WEMs can extract substantially more water per unit of time; further, because of their very high fixed costs, their owners are under economic compulsion to operate at high levels of capacity utilization.

As a result, modern WEMs often tend to interfere with other modern or traditional WEMs in their neighbourhood, reduce their water yields and increase their extraction costs. Such interference tends to increase as the population of modern WEMs in an area increases. Further, for the same discharge rate and extent of pumping, a tubewell which affords a deep drawdown will have a greater area of influence than an open well which acts as a storage reservoir especially when pumping is not constant. The drawdown caused by the same rate of pumping will be less in an open well which continues to get recharged slowly during the night when pumping is discontinued. Again, wells located at the centres of landholdings tend to have less overlapping areas of influence than those located at the boundaries. The extent to which wells are affected by interference will also depend upon the depth and the number of aquifers they themselves tap. A shallow open well will be hit hard: but if several tubewells interfere with each other each will impose a countervailing diseconomy on the others. Well interference is normally a temporary problem and assumes serious dimensions mainly because most members of the community tend to lift water during the same period. The impact may, however, be serious on owners of traditional wells when the monsoon fails.

The clustering modern WEMs in a small area and rapid increase in the cultivation of water-intensive crops such as sugarcane, banana, etc., often lead to a rate of water use higher than the rate of aquifer recharge: in such cases, the water table declines permanently and many shallow wells either suffer major decline in water yield or become permanently dry. Dhawan has examined several cases where mining of ground—water led to permanent decline in the water table. In parts of Mehsana district in Gujarat, over-exploitation of the aquifer pushes the water table down by 5-7 metres every year. Dhawan argues that the disappearance of traditional wells in Punjab can be partly explained by the lowering of groundwater tables far beyond the technically feasible limit which can be reached by traditional water lifts. In the hard rock areas of southern India where most wells exploit limited aquifers which are hydraulically unlinked with each other, digging a well faces a high risk of failure and, therefore, it is customary for neighbouring farmers to dig wells very close to a successful well as in many parts of Telangana region, Andhra Pradesh. This usually proves a very expensive method of sharing the limited potential offered by a small aquifer.

Well interference as well as permanent lowering of he water table have evoked serious concern among researchers and policy makers. Official response has usually taken three forms: (a) norms of spacing to be maintained between existing and new well; (b) state groundwater departments' categorization of areas into white, grey and dark; and (c) a licensing system of sorts operated with the help of public sector banks and SEBs. In

Gujarat, for example, a proposed new tubewell will not be allowed within the command area of a state tubewell or within a radius of 680m of an existing tubewell over 150 feet deep. The applicant is required to secure the consent of neighbouring WEM owners before an electricity connection or bank finance is provided.

The actual effect of such norms and the manner of their enforcement are often quite inequitable and regressive. First, since the spacing norms do not apply to a modern WEM being located close to a traditional WEM, they merely seek to protect resources-rich early exploiters from late exploiters; but do not offer any protection to existing owners of traditional WEMs who are usually poor.

Again, because of the large investment requirements of modern WEMs and because of their obvious appeal to farmers with large holdings, most early adopters of modern tubewell technology were large and medium farmers. Spacing regulations which have come in more recently thus serve to exclude the poor who are late entrants into the game. In other words, spacing regulations tend to single out the poor to bear the cost of maintaining the ecological balance.

In the 14 villages of Karimnagar district served by a highly successful multipurpose co-op, the resource-poor latecomers vehemently fought against the spacing norms until the co-op gave in and provided a large number of loans to small holder groups. Over the years, water tables have fallen in this as well as in other talukas. In the 14 Mulkanoor villages, the poor shared the prosperity and are now sharing the costs; but in the neighbouring talukas, the poor only shared the costs.

To add to this, since the norms are enforced through

banks and electricity boards, the well-off farmers who can finance their own investment and afford the somewhat costlier diesel engines remain completely untouched by them. Unofficial premia on electricity connections are common and often quite high. In Pandalaparru village, in a grey taluka of West Godavari district, such premia went up to Rs 10,000 per connection; but in parts of Gujarat, they were lower at Rs 2000.

The equity impact of the fall in water table and well-interference externalities is particularly adverse with autarky because in this case the entire irrigation surplus would be usurped by owners of modern WEMs who would maximize irrigation surplus per unit of own land by growing water-intensive crops. If in the process they render shallow wells using traditional WE technology dry, their prospective will have an immiserizing effect on the small farmers. In contrast, the emergence of water markets alters this scenario drastically. As we earlier discussed, if the community has competitive water markets, then who owns WEMs makes little differences to the distribution of irrigation surplus. The owners of traditional WEMs who suffer the external diseconomy would have the opportunity to buy irrigation at a price which is no higher than the real incremental cost of pumping. We examined in chapters 4 and 5 the evidence thrown up by the Shah and Raju study for West Godavari as well as Kheda villages where these theoretical results are amply supported by empirical data, and which confirms that the resource-poor would be better-off with competitive water markets even when they have to bear the brunt of the well interference externality.

Many existing public policies however militate against the emergence of competitive markets. Many electricity boards, for example, formally forbid electric WEMs to sell water under the contract signed before providing power connections. A state groundwater department would not permit private tubewells within the command of an STW, regardless of whether the STW works or not, or is able to irrigate its design command or not. Similarly, spacing and licensing norms, enforced by groundwater departments and NABARD create and strengthen the monopoly power of existing owners of modern WEMs who often can and do use this power to extract exorbitant prices for water from others. In many parts of Gujarat, the license for a new tubewell would not be provided unless it is at least 680 m away from an existing tubewell; the area monopolized by the existing tubewell—and in which late entrants have lost, for ever, their right to sink their own tubewell—can be as large as 500 acres or more. Early entrants thus have a great opportunity to become water-lords. In many parts of Gujarat, for instance, highly oligopolistic water markets have placed such water-lords in commanding positions in the village economy.

Saline ingress in coastal areas

Coastal areas have a very fragile groundwater balance. In coastal aquifers, fresh water with its lower specific gravity floats on a layer of saline water. If excessive pumping lowers the water table beyond a certain depth, this causes a difference in head which leads to the intrusion of sea water into the aquifer thereby rendering large areas of land unirrigatable with groundwater and often unfit for cultivation.

This form of diseconomy has assumed serious form in coastal areas of Tamil Nadu and Gujarat. Until 1970, the coastal areas of Saurashtra were agriculturally prosperous, full of mango orchards and known as *'Lili Nagher'*. In the mid 1960s, in its enthusiasm to promote a green revolution in the region, the government granted liberal loans for the installation of tubewells until a substantial proportion of farms had their own tubewells. The withdrawal of groundwater increased tenfold: where a traditional 'rahat' used to draw between two and four cubic metres in an hour, a tubewell could lift 20-40m/hour. Food crops were replaced on a large scale by sugarcane, encouraging rapid growth of sugar factories in the area. As a result, water tables along the coastal belt fell by 3-10 m over a period of seven or eight years. In 1970, most farmers in the area suffered reduced crop yields, and found well water brackish; some farmers continued to irrigate with the saline water thereby ruining their top sold. Successive droughts during 1972-74 led to further decline in the water table and more saline ingress. In some parts of the area, well water had 10,000 ppm salt content, much beyond the safe limit for drinking purposes and for irrigation purposes. Over 12,000 wells became saline in the small strip between Madhavpur and became saline in the small strip between 40 metric tonnes to 2 mt per ha; of coconut, from 22,000 units per ha to 8160 units per ha; of banana to less than a third, and of bajra and sugarcane to half, of their 1970 level. The market value of this prime agricultural land plummeted from Rs 60,000 to Rs 15,000-18,000 per ha. In several villages, half the families, or even more, including those of some large land owners, migrated in search of work and gave away or sold their animals. In fact, the unfettered exploitation of groundwater in the entire coastal strip stretching from Bhavnagar to Lakhpat imposed largely irrepairable damage on 1.2m, ha of land affecting some 1.3 million families in over 800 villages.

A study of the process of salinity ingress in the Minjur aquifer near Madras tells the same story. A 350-km strip of coastal farm land has been rendered largely unproductive due to saline intrusion caused by large scale pumping of groundwater over the last 35-year period. Most farmers of this once prize paddy land have had to switch to rainfed crops; many have sold off their lands at discounted rates. As in Saurashtra, land prices here too have crashed. Drinking water has emerged as a major problem and even tender coconut water has turned saline.

Coastal areas are possibly the only areas where the emergence of private water markets can exacerbate ecological disequilibrium. The major need in such areas is to contain pumping of groundwater to 'safe annual recharge', and the operation of water markets makes this more difficult to achieve particularly under flat tariffs which encourage WEM owners to pump more water than they would under metered tariffs. The present thrust of official policy in these areas is on 'hardware' solutions; more check dams, more tidal regulators, etc. But major long-term policy options here are; public monopoly over water, community management of the groundwater resources; import of water from external sources; and restriction on total pumpage through instruments such as quotas on power supply for pumping. We take a closer look at the dilemmas of a coastal community in Gujarat which has been at the receiving end of this diseconomy for close to a decade.

Behavioural foundation of ecological imbalance

Externalities associated with private development and exploitation of groundwater resources—and the environmental ill-effects they normally produce—are generally considered and analyzed from a macro perspective. The source of the problem, however, is micro and can be traced to characteristic behavioural patterns of farmers as economic agents. Neo-classical economics suggests that the free play of market mechanisms has the capability to self-correct certain types of distortions in the economic system. These equilibrating mechanisms, however, cannot adequately address some classes of problems. These are typically treated as cases of market failures. One major class of such problems relates to inequality in basic endowments of different groups in the economic system; the market system cannot correct these inequalities and in fact, may often reinforce them. An egalitarian economic system therefore requires public policy to check the free play of market forces. The second and equally important class of situations where market forces cannot automatically correct distortions are those characterized by externalities.

In case of the management of natural resources, externalities arising out of private development and use have been extensively studied by social scientists in recent decades. A major sources of interest in these issues is the essential commonalty in the variety of was through which private exploitation of natural resources results in ecological degradation in over-populated communities. Hardin's early paper provided the first and the simplest exposition of this process paraphrased as 'The Tragedy of the 'Commons', Subsequent works have focused on developing extensions of the 'Prisoner's Dilemma' framework for different natural resources under common property management regimes including common grazing and forest lands, groundwater resources, marine and inland fisheries, etc. Indeed, this framework has powerfully brought home the behavioural patterns associated with the assignment of common property rights in natural resources.

In a simple exposition of the problem, each user of a commonly owned resource seeks to expand his private gains by maximizing the use of the resources. As long as the number of such users is small compared to the rate at which the productivity of the resources gets replenished, the tragedy does not occur. However, as pressures of commercialization and population growth mount, common properly resources tend to degrade with over-use. In some situations, common property rights give way to some form of private property rights. In such cases, the behavioural processes of users undergo fundamental change. If the property right regime does not change, over-use occurs.

The Prisoner's Dilemma analogy used to illustrate this process involves two prisoners imprisoned for minor crimes involving short sentences. They are asked to confess to a major, jointly committed crime, without any opportunity to consult each other before replying. The number of years each would get as sentences in each possible case is given in the following table. The optimal joint outcome for both of them would result only if both do not confess. However, since neither is certain of what the other will do, each will be forced to pursue a strategy which will maximize his own gain on the basis of an assumed

Classical Prisoner's Dilemma

Prisoner I		
Prisoner II	Does not confess	Confesses
Does not confess	2/2	12/0
Confess	0/12	10/10

decision by the other. Prisoner I would reason that he would be better-off confessing, for, if Prisoner II also confesses, he would get 10 years instead of 12; and if Prisoner II does not confess, then, of course, he will get 0 which is even better. Prisoner II also reasons likewise. As a result, both confess and get 10 years each. Collectively, both are worse-off in relation to any of the other options open to them. The two prisoners will home in on the best option of two years each only if each has complete confidence that the other will not confess or if they are able to consult each other, and adopt a cooperative strategy. The analogy of the Prisoner's Dilemma with the users of the common property resource is clear; and so are the lessons. As a community, people will arrive at the optimal outcome if they are able to consult each other or if each has complete faith in the others.

Thus, at the micro level, the behavioural processes that result in the problem of saline ingress that we discussed in the last section are essentially similar to the Prisoner's Dilemma. Newspapers reports tend to focus only on the macro view of the calamity; they seldom highlight the fact that only a small fraction of the 1.3 million rural families of the Saurashtra coast hit by saline ingress were actually responsible for causing it. Each one of these was operating in a manner similar to the prisoners in the example above. Many experts believe that the ingress could have been controlled through some kind of regulation of ground-water withdrawal. Some went so far as to suggest that even individual villages could have minimized the damage to their land by controlling their own withdrawal from the wells falling within their own farm lands. However, others believe this to be impossible and argue that only control over withdrawals over the entire region can provide an effective answer.

This controversy notwithstanding, the fact remains that a Prisoner's Dilemma type result obtained, although the land which was affected was a private resources and the fact of whether groundwater is treated as a private or common property resources would not have changed the outcome. A public 'bad' was produced by the actions of individual well owners as a 'privately rational' response which made perfect sense to them as rational wealth maximizers.

This apparently suicidal behaviour can be understood only if we put ourselves in the shoes of a well owner who has struck it rich by irrigating three crops with well water and also selling water to neighbours. What kind of public policy stance would make sense would depend upon our assessment of how individual well-owners view their own actions vis-a-vis the community. There are three possibilities about how a well-owner relates to the potential crisis of saline ingress to which he, among others like him, is contributing.

(1) He is completely unaware of the impending doom to which he is contributing his full might. Innocent of the serious consequences of his actions, he continues to pump away at his tubewell as long as his private gain exceeds his private cost on the margin, completely obvious of the exceedingly high social cost that he imposes on the entire community by pumping his well. And when the doom becomes reality, be claims ignorance and acquits himself of the responsibility. If we assume that this is the general scenario, then the implication for public policy is to provide knowledge about the casual link between pumping and saline ingress.

(2) It is, however, highly unlikely that a tubewell owner in a coastal area is unaware that excessive pumping causes saline ingress, because it is happening all around him. But it is possible that a pumper may believe that, by some miracle, when the disaster comes, he will be spared while others are ruined. Or, more likely, he believes that he is but one of a large number of pumpers and that even if he exercises restraint, there is no reason why others would be as noble and also control their pumping. And, he reasons, if all the others do exercise restraint, then his uncontrolled pumping alone would not bring about doom. And therefore he goes on pumping.

(3) In the third situation, which is the most realistic, our pumpers is actually aware that his pumping is contributing to the impending doom; but he is also aware that disaster cannot be prevented even if he discontinues pumping his well altogether. He also assumes that others also think like he does, and calculates that doom is certain in, say, ten years. So he has only nine years to earn and after that he will have no means of livelihood. In this scenario, he begins pumping more vigorously than ever, as do other pumpers, under the same reasoning. As a consequence, instead of the tenth year—which is how long it would have taken art normal rates of pumping by all pumpers—the disaster actually falls in the third years. This is what we call 'competitive pumping' which has exacerbated the ecological disaster in coastal Saurashtra and in many hard rock areas of the south Indian peninsula. There is no doubt that this is the consequence of groundwater develop-ment at private initiative.

In terms of the prisoner's Dilemma framework, the dilemma of a pumper can be described as in the following table. We note that each pumper decides on his behavioural responses essentially by making assumptions about how others decide. If P is the net profit per well from restrained pumping at a level which is sustainable in the long run, and P* the additional gain that is possible through water-intensive technologies, then the ideal strategy for the entire coastal community is of restrained pumping by all which would ensure private return equal to P for all pumpers for all time to come. However, if only the ith pumper restrains pumping while others do not, he ends up 'a sucker'.

The dilemma of the pumpers in coastal Saurashtra

other farmers/farmer	Water saving crops	Water loving crops
Water saving crops	p/p for ever	p/p + p+ for ever
Water loving crops	p + p*/p now 0/0 after n years	p + p+/p + p+ now 0/0 after n years

After 10 years, when disaster strikes, he will get no reward for his ecological scruples. However, if the large majority of pumpers agree to restrained pumping, then a sustainable solution is possible even if a few prove difficult. The most likely outcome is of course that denoted by quarter IV of the above table which represents the tragedy of the commons; and the most desirable, by quarter 1, denoting restrained and sustainable use.

An exactly similar situation occurs in a nearly opposite ecological condition when the opening of access to cheap canal water raises the water table due to seepage and unlined channels; and canal irrigators begin switching to water loving cash crops such as banana, paddy and sugarcane. Here again, each farmer pursues privately rational strategies and, in fact, expands water use until the marginal contribution of water to his gross income per acre becomes very nearly zero, since in most canal projects in India water is charged at very low per acre rates. The Indian experience suggests that typically, within 6-10 years after a new canal opens up, large areas in the command begin to suffer from waterlogging and soil salinity. This is another classic case of the tragedy of the commons, but in reverse.

For each canal irrigator facing prospects of waterlogging and salinity in his land, low returns from crops using less water, such as food grains, for eternity would be preferable to high private returns from water loving crops for a limited period, say 5-7 years, after which the productivity of his land is certain to fall drastically. Most farmers in such a predicament are aware of this rationale. However, unless every farmer in the area decides to exercise restraint in water use on his own farm, it does not make economic sense for may one farmer to do so. On the contrary, it makes best sense for him to earn as much as he can while the land is still good. When all the farmers in an area work on this logic, they collectively hasten the common destruction of all land in the neighbourhood. In this case, as in the earlier one, the land being destroyed is private property; but water is perceived as a free good hence a public 'bad' is generated.

It is important to note that in this case also, one wise farmer trying to use water judiciously would not be able to do anything; as a matter of fact, he would end up a loser if he grows only foodgrains while all others use canal water to grow more profitable banana and sugarcane. Unless all or a majority of the farmers restrain water use, there is no way an individual farmer can save his land. The tragedy caused by excess water supply here is exactly the same as the tragedy caused by saline ingress in coastal areas, and both are analogous to the prisoner's Dilemma problem.

It is surprising that most research and action programmes focusing on ecological degradation caused by private use of natural resources have worked on solutions involving institutional mechanisms for joint action by the community, i.e. by all those involved in the situation. In canal commands as well as in coastal areas, a major option for sustainability is to create a platform for all the resource users to come together, to discuss and evolve norms for restraint on resources use. It seems unlikely that major answers to sustainability issues in natural resources will come from the institutional side, at least for quite some time to come. In the immediate future, it would be most functional and practical to approach sustainability issues in water resources from the angle of basin-or region-level water resources management strategies. This approach seems most practical because in most areas it is possible to plan, manage and operate water resources strategies at a level of aggregation where surface and groundwater resources can be used conjunctively to maximize their economic potential, while at the same time keeping a tight control over the ecological balance. In the remainder of this chapter we explore the basic potential for evolving and operating such strategies in a canal command.

The problems of waterlogging in canal commands

In the command areas of many of India's canal irrigation

projects, rising water tables and the build up of salts in the top soils have emerged as a major socio-ecological problem. Excessive canal water use and rapid rise in the cultivation of water loving crops by canal irrigators encouraged by low per hectare water charges, seepage from unlined distributaries and field channels, and inadequate provision for drainage facilities are identified as key contributory factors. Investments in drainage and canal lining supported by organizational reforms involving greater participation by users in maintenance and management of the systems have been widely discussed as a feasible solution. In this context, the importance of ground-water markets in raising the potential value of tubewells as irrigation-cum-drainage and pure lateral drainage mechanisms in canal commands has not been recognized by researchers and technologists. As a matter of fact, a series of tubewells along the banks of the Satluj-Yamuna canal in Haryana has been used effectively to contain potential rise in water table and to pump ground-water into the canal to augment supplies to the tail enders. Likewise, the Khairpur Irrigation project in Pakistan also uses public tubewells as pure drainage and drainage-cum-irrigation systems. Such a strategy of conjunctive water use depending on public tubewells, however, has inherent limitations. Public tubewell extraction capacities in a given region would tend to be insignificant when compared to those available with private pumpers. Then, there are problems of maintenance and proper management of public tubewells which severely impair their performance. The Khairpur project worked well for the first ten years but later suffered from lack of maintenance. For large scale and long-term impact on conjunctive use therefore, it is important to find ways to induce private pumpers to operate then pumping plants as irrigation-cum-drainage systems. The main difficulty here is that their private returns from pumping in high water table areas are lower than the long-term returns to the community as a whole, a structure of incentives has to be devised that would match private incentive with the needs of the community. This is the classic example of a situation where hydro-geological factors interact with socio-economic and institutional variables.

Figure 1 is divided into two parts: the upper half, which concentrates on the economic aspects, describes the relation-ship between ground and canal water use and the incremental private and social costs and returns attendant to such use. To the extent that private costs imply allocation of real resources, they are costs to society as well; but social costs exceed private costs when private action imposes diseconomies on others or on future generations. They are less than private costs if private action saves real resources needed to mitigate an undesirable outcome. In the lower half we examine the relationship between ground and canal water use and the movements in the depth of the water table. To simplify the analysis, we define d_1, d_2 as the safe range within which society desires to contain the movement of the groundwater table; a fall in the water table below d_2 implies dangers of aquifer mining; a rise in it above d_1 indicates dangers of waterlogging and soil salinity; under traditional WE technology, the water table keeps moving within a narrow range about d_0. To start with, we concentrate on the right half of the figure which shows that the opening up of modern WE technology creates possibilities of highly profitable irrigated agriculture as reflected in the large area $rc_{11}e_1$ enclosed by the private return (g) and private cost (c) curves. Private users will

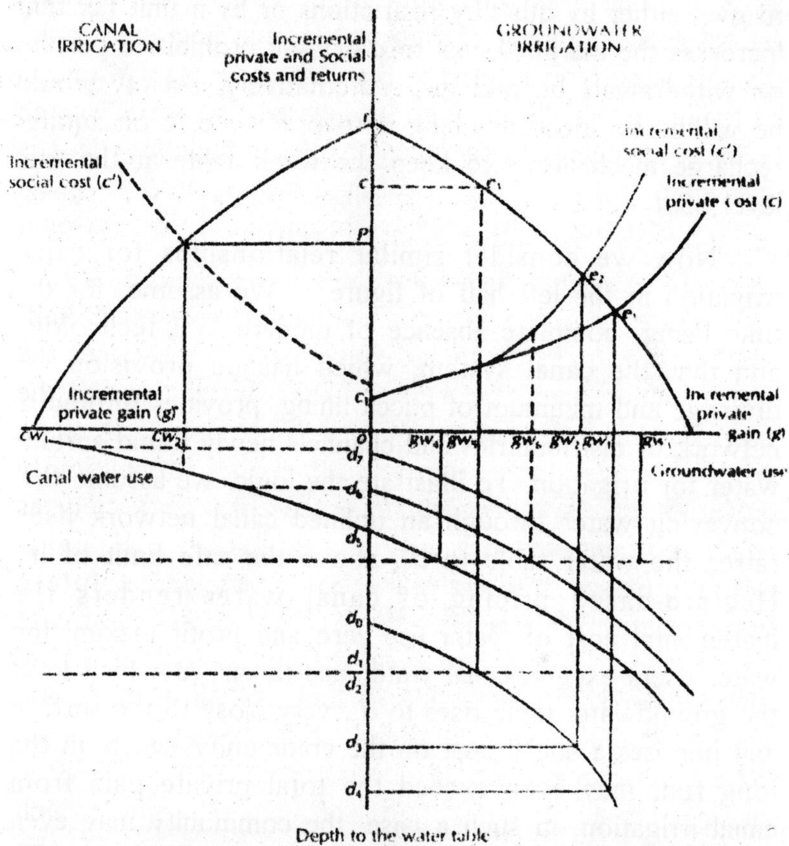

The logic of conjunctive of ground and surface water

expand groundwater withdrawals to gw$_1$ which, however, will force the water table down blow d_2 to d_4 and impose social costs along with incremental social cost (c) curve. A marginal tax on water users which discounts adverse future outcomes at a finite interest rate will reduce water use to gw_2 and force up the water table to d_3 which is still below d_2 and will result in progressive depletion of the aquifer over a long time period. The diseconomy can be completely eliminated only by containing groundwater use to gw_5, either by quantity restrictions or by a unit tax that increases the marginal cost to c or by a prohibitive penalty on withdrawals beyond gw_3. Alternatively, society would be willing to spend anything upto $c_1\ e_1\ e_2\ e_3$ c on aquifer recharge mechanisms to keep the water table at the safe level of d_2..

Now we consider similar relationships for canal irrigation in the left half of figure 1. We assume, for the time being, complete absence of modern WE technology and that the canal system, which has no provision for drainage and minimum of pucca lining, provides through a network of distributaries and channels newly found surface water for irrigation. To illustrate the logic, we assume that conveying water through an unlined canal network itself raises the water table to d_5, above the safe limit of d_1. Hectare-based pricing of canal water renders the incremental cost of water use zero and profit maximizing water users expand canal water use to cw_1 at which level the groundwater table rises to d_7, very close to the surface and imposes a social cost on the community which, in the long run, may even exceed the total private gain from canal irrigation; in such a case, the community may even be better-off without a canal system. Alternatively, volumetric pricing of canal water at P will contain water use at cw_2, and restrict the rise in the water table to d_6.. We note that even completely abolishing canal irrigation will not eliminate the social cost of waterlogging under our assumptions.

We note that if the canal system is used only as an aquifer recharge mechanism and not for irrigation, then not only does the social cost of ground—water irrigation with modern WE technology become zero until groundwater use expands to between gw_4 and gw_5, but the social cost imposed by the canal system in terms of water-logging, too, becomes zero. It is also evident that as canal water use increases, there would be a countervailing need to increase groundwater withdrawals. In sum, then, an optimal conjunctive water use strategy in canal commands implies encouraging groundwater irrigation in precisely those areas where incentives for it are likely to be minimum; where cheap canal water is available aplenty. It would also imply using the canal network more as an aquifer recharge mechanism in such areas as a long-distance water transport system so that tail enders get more canal water and are discouraged from excessive exploitation of the aquifers away from the canal head where the rate of recharge is lower.

Empirical evidence, though limited, indicates that the actual build-up of private water extraction capacity in canal commands follows a pattern opposite to what is required to ensure a stable equilibrium in the water table movements. An econometric investigation of 20 years' data of monsoonal and inter-monsoon changes in the water table over some 100 locations in the command area of the Mahi Right Bank Canal system in Gujarat indicated that while rainfall precipitation and canal water releases for kharif crops were the main determinants of the monsoonal rise in water tables in different areas, differential rates of build-up of water extraction capacity caused water tables to fall at varying rates in different locations during winter and summer months. Thus, while the capacity of private pumping to lower water tables in water-logged areas is clearly demonstrated in actuality it does not help to stabilize water tables because the build-up of private we capacity tends to be low in areas with very high or very low water tables.

In sum, thus, while pumping groundwater can cause ecological havoc in areas suffering from saline ingress or impose high social costs in hard-rock areas or in peripheries of canal commands, it can confer larger social benefits in water-logged areas of canal commands. Basin-level transport of water from the second type of area to the first thus offers a major long-term option for sustainable water resources management. But a more immediate and implemetable options is to work through private pumpers.

Limits of existing policies

A variety of instruments used by public policy makers in various states for sustainable water resources management have failed to produce viable solutions in any of the three situations discussed above. Each instrument has been used in a piecemeal and *ad-hoc* manner; while some have produced limited results, others have either failed or run counter to what should be important public policy goals. Licensing and spacing norms, as we saw, have been difficult to enforce when they defy immediate private interests. It might be argued that this effective defiance has, in balance, been all to the good since the regulations themselves are likely to have had retrograde effects. Likewise, in waterlogged areas of canal commands, the command area administrations have found it impossible to contain over irrigation and wasteful use of canal water in the face of acre-based water pricing policy. In the coastal areas of Saurashtra, even as the irrigation department has begun to spend large amounts on check dams and tidal regulators to prevent salinity ingress, private farmers continue to pump their wells while at the same time using subsidized gypsum, which is beginning to have a progressively declining leaching effect.

A comprehensive land-water policy framework which may effectively respond to these complex problems would recognize the importance of balancing the various externality and equity effects as special cases of the larger task of achieving optimality in spatial, interpersonal and intertemporal use of water resources. Among other things, such framework should aim at:

(a) exploiting to safe limits all water resources available in a given region and ensuring, to the extent possible, equitable access for all;

(b) minimizing inter-farmer and inter-generational diseconomies in exploiting and using water;

(c) devising workable ways of minimizing groundwater withdrawals in coastal areas; and

(d) promoting conjunctive water use where both canal and groundwater are abundant.

Past experience in dealing with this task in many countries indicates that legal, quasi legal and organizational instruments of public policy will not, on their own, succeed in securing the compliances of farmers unless they are accompanied by measures aimed at affecting private returns to irrigation in different regions, or unless the structure of property rights on the water resources itself is drastically reformed.

Public control of groundwater resources

In theory, effective public control over all water resources could neutralize all the effects discussed earlier. In such a scenario all private WEMs, traditional and modern, would be delicensed; a public authority, assuming the role of a monopoly extractor of groundwater would locate a small number of modern WEMs in each farming community and provide water to all users at a price equal to the long-run marginal cost of extraction. Where there exists danger of over-exploitation, the authority would charge an additional premium to contain the rate of water use to the rate of recharge and use a proportion of the surpluses so generated to explore feasible ways of improving the rate of recharge. From the viewpoint of equity, an ideal state could be established if the profits earned by the monopoly are distributed equally among all the members of the community including the landless. The questions of optimal use and equity will thus get resolved simultaneously.

Since all the externalities associated with private exploitation arise primarily because losers find it impossible to extract suitable compensation from the emitters of the externalities under the existing structure of property rights, public control over water resources by a well-informed and just authority will result in their effective elimination. When private WEMs do not exist, well interference will not be a problem any longer. Likewise, in areas of water stress, the public monopoly will be able to restrict water use to safe levels through pricing policy and, if necessary, through quantity restrictions. In canal commands, similarly, it would ban altogether or restrict canal water use near the head and along the canals where water tables are likely to rise due to seepage. By maintaining parity between canal and well water, it could, in theory, eliminate the existing preference for cheap canal water over groundwater. Thus, in addition to neutralizing all the harmful effects discussed earlier and harnessing the beneficial effects, effective public controls of all water resources could substantially reduce society's investments in WEMs and drainage systems.

In spite of all these highly desirable features, few researchers pay serious attention to the alternative of public ownership and control of water resources. For one, there is not guarantee that the local-level bureaucracy under public control will either be effective or demonstrate the sensitivity needed to appreciate the complex questions of equity involved. World-wide experience with public systems managing major and even small irrigation systems has been so uniformly disappointing that most reasonable viewpoints now increasingly favour decentraliza—tion of management and control of water resources to farmers and farmer groups. The review of a relatively rare example of such an effort in Pakistan showed that the entire public programme suffered from a variety of technical and administrative problems; but above all, the review concluded that 'the programme pre-supposed the existence of a management structure and skills that did no exist in the irrigation bureaucracy at that time' Likewise, while small farmer groups are known to have achieved considerable success in neutralizing localized externalities and inequities by reducing transaction and bargaining costs, it cannot be safely supposed that such groups will be any more considerate than individual water users to other groups with whom they cannot directly relate but who might be affected by their actions.

The effectiveness of existing policy instruments in neutralizing the externalities: Gujarat

Policy instrument groundwater	Situation1: equity in the access to exploitation	Situation2: well interference and over diseconomies	Situation3: saline ingress externalities	Situation4 conjunctive water use
1. Well licensing policy	-	?	+weak	-strong
2. Spacing regulations	-	?	+weak	-
3. Public tubewell water supplied at cost or subsidized rated	+	+	-	+strong
4. Cheap finance to the resource-poor for WEMs	+strong	-	-strong	+strong
5. Power pricing system				
(a) pro rata	-	?	+	-
(b) flat rate per hp	+	?	-	+
6. Canal water supplies at low cost	+	+	+strong	-strong

3

Irrigation Development: Implications of Recent Experience for Aid Policy

Rational for irrigation

Irrigation has been a favoured sub-sector for public, private and aid-donor investment. The world Bank estimates that there is now 160 million hectares of irrigated land in the developing world. Irrigation now serves 20 per cent of all harvested land but it receives 60 per cent of all fertiliser and produces 40 per cent of all crops. The investment in irrigation exceeds $15 billion and it is growing at more than two per cent per year. It is an approved area for aid donors receiving nearly one-fifth of all aid for food and agriculture.

Irrigation is the principal means by which man modifies climates to increase food supplies. New developments in irrigation technology plus complementary advances in plant breeding, crop protection and agronomy "packages" have increased the potential productivity and profitability of irrigation agriculture. This increased productivity comes from higher yields, multiple cropping, often two or even three crops a year, and reduced risk of crop failure. Therefore it is argued, mainly by those with a technical bias, that donor interest and public and private investment are well-founded. Without irrigation the "Green Revolution" would founder.

But irrigation investment also has had loud critics who note such matters as the huge costs involved, costs which are often underestimated, the delays in construction, the yields below forecast, the poor financial performance, and the evironmental damage to soils or human health. All too often the promise has been greater than performance.

Irrigation and rainfed agriculture

The prime motivation for promoting irrigated as opposed to rainfed farming is technical. It is theoretically sound and a well verified fact that rainfed farming is a high-risk low-productivity system of agriculture and irrigation is a low-risk, high-productivity system. The paradox is that in practice rainfed farming is often economically efficient whereas irrigation farming can be economically disappointing. It is a puzzle as to how it is that a system of agriculture which can produce average countryside yields of say 5 to 6 tons of paddy per hectare is inefficient compared to a system under which it would be difficult to exceed the 2-3 tons per hectare average.

The answer to this paradox lies in the financial and human resource scarcity. A very heavy investment is required to produce irrigation facilities, skilled management and further financial resources are needed to operate them effectively, and higher levels of input services and farmer accomplishment are also demanded.

Irrigation and rainfed arming are not always competing for resources. IFPRT Report on the sources of additional food for the 1990s argued that, although irrigation was likely to be the dominant source of incremental food supplies in Asia and North Africa and the Middle East, it was relatively unimportant in Latin America and almost negligible in sub-Saharan Africa. Such continental aggregates hide facts of vital national and local importance. For example, rainfed agriculture is forecast to be much more significant than irrigation in Burma and as important in Indonesia, Nepal, the philippines and Sudan. Irrigation will make a material contribution to food production in some sub-Saharan countries such as Madagascar, Mali, Senegal, Cameroon, Ivory Coast and Nigeria.

It would be a false perception to see rainfed and irrigated agriculture as always major competitors for scarce resources. The main competition is between agricultural and non-agricultural investment. But as will be stressed in this paper there are hidden dangers to aid donors in irrigation including its:

— apparent non-controversial natural;

— simple conception, difficult execution;

— various areas of ignorance: ignorance on technical, social and organisational matters.

Once agriculture receives due acknowledgement in irrigation planning and management it will still be necessary to temper the technical enthusiasm of the professionals with economic discipline. Agriculture shares with engineering a tradition of technical excellence. Obtaining maximum yield per acre, meeting potential evapo-transpiration, and ensuring sowing, transplanting, spraying and so forth at the correct time as revealed by experiments, are all criteria by which agronomists judge performance. But in the practical world where numerous resources are employed in an integrated complex system, each with an alternative use, it is often necessary to judge success vary broadly and to recognise merit in "second best". There is often a wide gulf between maximum yield and optimum yield, that is between technical and economic efficiency, and despite the attractions of the former it is economic efficiency for which we have to plan. Agriculturists' preoccupation with high productivity may limit a broader vision that includes the opportunity costs of obtaining the yield and the social consequences of the uneven access within a country to the irrigation facilities, and in the longer term the environmental impact of short-term productivity goalds.

More food and more stable food

Whilst it would be difficult to imagine irrigation failing to increase yield per hectare of cereals it is less certain that it stabilises production. Although conventional wisdom is that modern technology is yield-stabilising, Barker et al, 1981, following a rigorous empirical analysis, contend that, at least for rice, absolute variance in yield increases with irrigation. This occurs because of irrigation's interaction with other high-yielding technology such as fertiliser which is a destabilising technology. They note that relative variability may also be increased if more intensive agriculture, fewer varieties and so forth increase crop losses due to pest and disease attack. If irrigation is poorly managed or maintained, if the supply is extended beyond the water resource capacity, then unreliable irrigation will ensue. Unreliable irrigation supply from surface and groundwater is probably the major cause of variation in yield. More importantly, in the medium-and long-term, unreliability causes farmers to cut investment in high-value crops, fertilizers, sprays and so forth and is a major contributor to disappointing irrigation performance.

Modern agricultural technology is going to be increasingly used in developing countries. If modern technology, including irrigation, is to increase instability in yield then this has important policy implications. Aid donors can assist by promoting research to confirm the relationship. If it is found that irrigation, particularly in the early years, increases yield instability, they should not of course cease to support the modernisation process. Donors should aid the process but also promote the development of transportation, storage facilities, buffer stocks and other facilities which can reduce any problems resulting from yield variability.

Source of production benefit

Irrigation properly executed can transform the production potential of agriculture irrigation. It can:

— increase area cropped;

— promote intercropping and multiple cropping;

— ease changes in cropping pattern to more profitable crops;

— increase yields by avoiding moisture stress;

— facilitate profitable use of other inputs;

— reduce risks and encourage on-farm investment.

In arid countries such as those in North Africa, and the Middle East through to Pakistan, rainfed agriculture is scarcely possible. In monsoon countries with as single rainy season supplementary irrigation can extend the season and encourage a second or even a third crop. The shift from single to multiple cropping is likely to be a more important source of additional food in the future than increases in area cropped. The other major source of benefit will be the encouragement of the adoption of other inputs. Joint application of a "package" of modern agricultural practices such as improved seed, fertilizer, crop protection and land management produces a complementary or more than additive production response compared to adopting of any single innovation. Reliable, timely and well-managed water supply and use is one of the most crucial components of a high-yielding agriculture. A reduction in the risk of loss through drought is an important determinant of the form and level of farmer investment including his own labour. It is of course all too easy to confuse technical with economic and social efficiency. The bright vision of transforming a desert to green pasture or to highly productive multiple-cropped farm land though technically feasible should no longer blind anyone to the economic and social problems involved in this transformation. The problems of irrigation set out later in this paper are formidable. Nevertheless, the potential benefits are real and probably represent the only prospect for food security for a large proportion of the inhabitants of the developing world.

Opportunity cost of water

In the earliest stages of agricultural development labour is the scarce factor of production. As population grows, first land and then capital become the limiting resources. In many countries we have now moved into an era where water scarcity is a dominant consideration. The taking of water to irrigate new land in the desert implies that the water is not available on existing developed land. The demands of urban areas for hydropower may conflict with ideal water storage releases for irrigation. Competition comes from the rapidly growing needs of urban and rural areas for plantiful secure drinking water supplies. In addition there are water interests in navigation, fishing resources and maintenance of river flows for pollution control and preventing saline water intrusion from the sea.

Use of the remaining surface and groundwater resources has high and increasing opportunity costs. Given that most water resource projects and water allocations are irreversible, great care and vision are necessary in current planning. It will be argued later that presently unallocated water resources may be a crucial "last chance" means of obtaining greater equity or fairness in the distribution of irrigation benefits.

Can irrigation help the poor?

There can be little doubt that public support for irrigation can help poor farmers but much more uncertainty as to whether it can directly assist the poorest who are likely to be the landless, the unemployed, the elderly, or single females with children. Perhaps what the poor most require is a reliable cheap food supply and a good irrigation scheme can help provide this.

Critics of irrigation argue with some ground that economically successful irrigation is rare. Furthermore even if it is successful the benefits to irrigation farmers have to be offset against the depressing effect that this can have on prices to rainfed farmers. This was part of the argument of those who were concerned with the "second generation" effects of the Green Revolution. However it is an argument for further research on technology and special policies for rainfed agriculture and not an argument to stem the progress of productive irrigation agriculture.

To help small farmers within an irrigation network the main requirement is that the design should be equitable and the scheme operated according to design. The normal concept of an equitable supply is a given share and timing of water in proportion to land owned. There are in existence some imaginative schemes for giving a larger proportional share to small farmers. For example, in Haryana India at Sukhomajri Project a water coupon scheme is operating giving water rights to all including the landless. A very dry area of Maharastra near Pune, the Gram Gouray Paratisthan, has a well subsidy scheme which gives well size and hence water in line with family size. Such brave and imaginative innovations deserve study support and, if successful, replication. Donor assistance can play a part in this process as many of the leaders and bureaucrats supporting these new ideas face considerable pressure from reactionary vested interests.

More conventional donor support can encourage resource development to the poorest. World Bank\United Kingdom-assisted rural credit to India carries the "small farmer only" tag. With experience it is proving possible to add to this requirement a suggestion that certain categories of investments have greater "reachdown" to the poor. Minor irrigation (shallow and deep tubewells, dugwells, pumpsets, electrification, land development, conservation and reclamation) has reasonable reachdown but not so high as animal investments. Plantations, forests, marine fisheries, market yards, and tractors have low reachdown. By analysis of the impact of various categories of credit, together with the scope for misuse, the real economic benefit, and the risks of failure and default, appropriate policies can be devised and evaluated. Similarly within the irrigation sub-sector it is possible to analyse the effects upon income distribution of alternatively irrigation strategies. The insights obtained should be tested for their general validity. The ways in which alternative ways of improving irrigation affect the poorest are not obvious and need study. Dozing et al reporting from the Philippines indicated that rehabilitation of small-scale irrigation schemes in the philippines benefitted the poorer labourers relatively more than the farm operators or landowners.

Good management of existing schemes is undoubtedly the best way to help poor farmers. It is becoming increasingly clear that sub-standard management of operation and maintenance is widespread and that under such conditions it is the poorest who suffer most. One can go further and assert that poor O&M of irrigation can be a major cause of impoverishment of those close to the margin of subsistence.

Problems such as this can be minimised if there is good diagnosis based on field research, good planning of schemes and highly professional management of operation and maintenance. Unfortunately, as evidence presented later will reveal, these pre-conditions are seldom satisfied. At a minimum promoters of irrigation have a responsibility to ensure that public finance should not be used to exaggerate income differences between different groups.

Rainfed agriculture or irrigation in Africa?

In most of Africa irrigation is either largely unimportant or unsuccessful. It is important in Sudan and Madagascar and significant in Mali, Senegal and to a much lesser extent in Cameroon, Ivory Coast, Nigeria, Ethiopia, Mozambique, Somalia and Zimbabwe. The temptation to see irrigation as the solution to arid land and unreliable climate should be resisted. The very depressing 1980 Club du Sahel Report— The development of irrigated agriculture in the Sahel— conclude that irrigation farming is not making an efficient contribution to food production; it is expensive to build. ($5 000-$20 000\hectares 1979 prices) and maintain; it is not obtaining high yields and double cropping; it suffers technical problems, management and training problems, agricultural policy problems and financing problems. Even the rehabilitation programme is seriously behind schedule. The more comprehensive 1981 report from the world Bank also urges caution; it notes that, despite considerable investment in African irrigation in the 1970s, there was hardly any increase in cultivated area in a number of countries because of land abandonment or rehabilitation needs. Not all developed areas are farmed and not all farmed areas are harvested. In most cases yields have stagnated or fallen (with some notable exceptions in sugar and rice) and in addition to the expected technical problems of management, lack of levelling and so forth, there is often an unfavourable price policy giving poor economic incentives to farmers.

Rehabilitation should be the first priority. The World Bank Report concludes that the following measures are necessary

- improving the operation of existing projects with respect to water management and agricultural services, and closer study of soil conditions;
- improving economic incentives for farmers and increasing their participation in operating and maintaining the irrigation schemes;
- rehabilitating infrastructure (drainage, water distribution, land levelling);

— reducing the size of holdings where they are too large to be cultivated intensively, and introducing double-cropping where feasible; and

— increasing the rate of cost recovery.

This will require finance and technical assistance. In the opinion of the writer, until most Sub-Saharan African countries demonstrate a capacity to operate their very small irrigation projects, the ambitious large-scale basin developments on the Senegal, Niger, Kagera, Badhera, various northern Nigerian rivers and Lake Chad basin should only receive an amber light. New developments should proceed when the technical and institutional basis is formed and proven. The donor community can assist in preparing these foundations by external technical and financial assistance. However, the simplistic optimism of the Brandt Report which talks of vast unexploited potential of large international river basins, needs treating with some scepticism, at least in Africa. In contrast to the large schemes, local, very small-scales, swamp irrigation and other water management projects deserve more attention. It is in Sub-Saharan Africa that links between minor irrigation and drinking water investment and soil conservation are worth further exploration.

The underexploited river basins in Asia—the Ganges-Brahmaputra, Irrawaddy and Mekong—have, as Brandt suggests, considerable potential, but it is no accident that they are presently under-developed. Taming these large rivers which are very prone to flooding is technically difficult and very expensive. There are also likely to be formidable diplomatic difficulties in obtaining agreement of all riparian states to a river basin development and management plan. Nevertheless, the potential is there and donors should explore development options.

Major problems encountered in irrigation projects

First a note of caution practically all development Projects are troubled by obstacles to efficient implementation. Rural development is particularly problematical. The following catalogue of problems associated with irrigation should not be interpreted as a unique set in kind or degree. Many of the barriers which will be cited, such as shortages of recurrent finance or inefficient administrative procedures in such matters as tendering, are common to such schemes as livestock projects, rural feeder roads and seed multiplication schemes. Other problems that will be enumerated, which appear to be solely irrigation problems, such as inefficient canal linings or lack of cross regulators in the canals, are in effect particular examples of general problems in rural development: that is, defects in design and\or construction and incomplete specification of projects.

Success and failure are not simple binary concepts. In assessing development there is an infinite gradation from unconditional success to abject failure. Furthermore, a wide variety of criteria can be applied given the multiple objective nature of the development process. Without irrigation the developing the developing world would be a very much poorer and certainly a more hungry place. Irrigation has undoubtedly contributed enormously to keeping per capita food supplies ahead of population growth. Nevertheless, barriers to adequate, reliable and timely irrigation do arise at all stages of the project cycle. Studies of operating schemes reveal that faults arise in pre-planning, project identification and detailed feasibility planning, construction and operation phases. This sequence is examined in the following review.

Pre-planning activities

Any action to build new irrigation or to extend or improve existing schemes requires knowledge of previous experience and basic data on resource availability. Often minimum information is absent, inaccessible or in an unprocessed form. We do not urge a purist approach and in fact commend the concept of "optimal ignorance" given the high cost of data collection. However, in irrigation the real damage caused by misinformation is high and often irreversible.

Basic concerns of planners such as sub-optimal operating performance are poorly understood. The extent of sub-optimality, the nature and influence of various causal factors, let alone the feasibility and costs of reform, are also largely unknown. Data are seldom in the right form because there are substantial definition and measurement problems. Even primary data such as the "area irrigation" statistics are unreliable. Recent agronomic advances which facilitate intercropping and multiple cropping of up to three crops a year are one source of confusion. Secondly, supplementary irrigation of rainfed crops is increasingly economic as the use of complementary inputs (fertilizer, crop protection and so forth) expands. Thirdly, there are definitional problems related to whether or not roads, canals, drains, field bunds and the like are included or excluded; the discrepancy can reach 30 per cent. The area statistics can be made even more unreliable by the deliberate distortions of records by those responsible for collecting the statistics. It is asserted that the area may be under-estimated when land revenue or other taxes are to be levied, or where the "missing" irrigation water can be illegally sold. It may be over-estimated when development grants are available on an area basis or where performance of extension workers or irrigation personnel is judged in part by area criteria. There is now some empirical basis for these widely assumed practices.

If area irrigation is misunderstood or misrepresented, it is hardly surprising that there is even more doubt about vital and complex operating performance data, such as canal discharge over time (say a day). Efficient management of an irrigation system, without the basic performance data is virtually impossible.

Physical data concerning soils, rainfall, river and canal discharge are examples of essential information for planning and operation. Not only are there gaps in coverage but there is evidence of deterioration. There are many reasons for this: decisions to hold inflation by cash limits on recurrent budgets which can affect field supervision; failure to increases the recurrent budget in line with development expenditure; trimming non-salary elements of recurrent budget; decisions to decentralise without sufficient experienced staff and equipment; increased political influence in technical matters; high priority to project planning as opposed to routine data work; premature change to computer data banks, are all examples known to the writer of reasons for a deterioration in basic data services. Donors can help the data problem through technical assistance for basic data collection on such matters as watershed management, land-use planning.

One reason for poor performance of irrigation is our level of ignorance about crucial components of the operating system. For example, we do not have good insight into natural-science questions such as the precise nature of plant-water-soil interactions in the field, into the effect of tropical climate (particularly extremes) and the impact of various levels of watertable or salinity upon crop yield. Once the social dimension or the system is incorporated—the water allocation patterns of farmers, their reaction to climatic variation and the influence of their behavior on soil salinity—our ability to predict falls still further. Attempts to engineer or manage the full water resource system are additionally fraught with problems because of our ignorance of the relationship between system components. Recourse top modelling, the normal engineering solution to complex problems, is unhelpful as specification, validation and testing are at present impossible. Indeed, resorting to impressive analytical techniques may be dangerous unless the limitations are clear to those interpreting the outcomes. What modelling exercises can usefully do in these circumstances is to assist understanding of the system, the components and relationships, help identify gaps in knowledge and guide research priorities.

Aid donors can assist in knowledge acquisition in ministries, public development agencies and local universities. Applied research activities as well as being essential to education and training, are overheads to planning and all too often not considered of first priority by governments who have numerous, apparently more pressing priorities.

National policy planning reports

Scarce water requires wise allocation and good planning requires accurate diagnosis based on field contacts. Such truisms unfortunately do not lead to obvious actions. Planning resources are scarce with high opportunity costs.

One rapid and cost-effective approach to reappraising national policies in this field which may provide a model are the recent experiences of WHO in assessing drinking water needs. In advance of the United National Water and Sanitation decade 1981-1990, WHO proposed a three-phase planning process to reassess and prepare national policies and project plans:

Phase I Rapid Assessment Reports

Phase II National Plans and Policy Review

Phase III Project plans.

This has been largely successful. The rapid assessment reports typically took about three man-months of senior professional time and were in effect mini-sector reviews.

In many countries there is now a clear case for a similar exercise for irrigation, possibly as part of a broad water resources assessment taking into account competing uses of surface and groundwater, especially urban and rural water supply, hydro-electric power and flood protection. If the second phase of preparing a National Water Plan and Policy Review is undertaken (say 30 man-months for a medium-sized country) then multiple interests, including the often neglected health and environmental issues, need to be confronted. The World Bank has taken a lead in this and there are some good prototype studies such as the recent Egyptian Irrigation and Land Reclamation Sub-Sector Review. Resources to undertake rapid assessment reports, plus a contribution toward sector reviews to ensure comprehensive coverage and where possible comparable experience, could usefully be provided by the aid donor community. The often overlooked area of policy assessment to provide the framework within which the more favoured project planning activities can flourish also needs more attention. It is worth noting that a national water policy and plan does not imply a uniform approach. Sector plans are a precondition for forward aid programming within or outside the framework of comprehensive and integrated country programming. Considerable variation in local conditions—environmental, cultural, institutional, technical and economic amongst them—dictate that locally sensitive strategies are included.

Capital aid

Treasury ministers in developing countries know that irrigation is a relatively assured source of aid and credit. However, capital aid comes in many forms and recipients are well advised to question the comprehensive nature of the finance (will it, together with local resources, cover all structures, drainage, transportation facilities, etc.?), whether local costs are met, whether the aid is tied to donor exports, what delays in disbursement are likely, whether the donor agency has the right to waive regulations and procedures should circumstances change. Water resource develop—ment is one area where it pays a recipient to "shop around" looking for the best terms and conditions. The corollary of this is that in many countries too many donors are chasing projects in the "safe" water sector. The writer worked in an East African Ministry of water Development where 18 bilateral and multilateral official agencies were giving financial assistance. Some of the most talented local planning and administrative personnel were tied up greeting, meeting and generally satisfying donor curiosity, whims, regulations and performance criteria. The care and attention given to visiting missions contrasted starkly with the cursory treatment given to the domestic annual budgeting process which may, in part, have accounted for the large shortfall in the recurrent budget and sub-optimal project operating performance.

The aid community should re-examine the problem of excessive donor "competition" for projects and attempt to find new forms of collaboration.

Local and recurrent cost financing

Attempts during the 1970s to bring a poverty focus to the selection of aid projects and the growing capacity of some developing countries, led by India, to supply almost all project inputs, highlighted the issue of local costs. It is now five years since DAC Guidelines on local cost. Financing were agreed and slow progress has been made in adjusting donor procedures in line with them. Much more could be done to increase the flexibility of aid disbursements. Local cost reform is limited by several factors but aid-typing arguments are most important and the latter could lead to less local cost finance in future.

In some donor countries, conflicts between interest groups reduce local cost contributions. For example, in the United Kingdom in the 1978\1979 to 1981\1982 period the ceiling on local cost for project aid to Asia countries was set at 38 per cent. In practice, the Treasury has a veto on the use of local cost on individual projects and so actual disbursement is nowhere near the theoretical maxima. In 1977, only 6 per cent of project aid was local costs. Unless this experience is atypical, it would appear that DAC Members are failing to realise the positive contribution, which they recognised back in 1977, that local cost financing could make to building local production capacity, to providing employment and direct, immediate income increases, and to broadening the range of programmes and projects eligible for aid. Since 1977, with a slow-down in economic growth, local funds have become even more scarce and it is likely that socially and economically good projects and appropriate technologies (e.g. labour-intensive construction techniques) are not obtaining backing.

In 1977 it was anticipated that projects to satisfy basic human needs and to promote self-reliance would receive increased local cost support. With a change in mood away from "provision" of basic needs to meeting these needs through directly productive projects. (projects such as small-scale irrigation), we should have seen an increase in local cost finance for such projects. The official aid agency response to local cost financing opportunities has been disappointing.

Recurrent costs are largely local costs for operation and maintenance purposes. The DAC guidelines on assisting recurrent cost financing are now three years old. Despite the undoubted fact that support for O & M is normally the most productive use of finance, given the weight of sunk capital costs in total costs, donor support for the recurrent costs is relatively rare. The meeting is asked to review the case for O & M support, relate to ways in which aid agency assistance could be granted for the recurrent budget, helping to overcome the apparent reluctance of many donors to provide this efficient form of aid.

The prime rule of aid planning should be not to make conditions of the target group worse Project planning is increasingly adopting social cost-benefit concepts and procedures which paradoxically may be creating financial problems and reducing the volume of aid transfers. Social cost-benefit procedures force analysts to create a "shadow" world where costs and benefits are in line with opportunity costs rather than market prices which for various reasons are distorted. Projects designed using these shadow-pricing procedures, which are mostly produced by and advocated by aid donors, will be labour-intensive (shadow wage rates lower than market rates) and will use less foreign exchange (foreign exchange is undervalued by the market). If high rates of discount are also adopted in line with notional rates or return this will encourage designs that are recurrent-cost-intensive. Under these procedures the type of project that is designed is labour-intensive in construction and operation phases and uses maximum of local materials. If the latest procedures are used which give additional weight to benefits that accrue to the poor ("social prices") then this is reinforced. If aid agencies advocate shadow pricing and high rates of discount that when translated into market prices increase local cost and recurrent cost components then it is surely incumbent upon more of the donors to relax their stance on local and recurrent costs.

Indeed, they can also help with finance and technical assistance to bring some operational reality to the shadow world by assisting recurrent cost financing. Social cost-benefit analysis is costly in time and human resources. Irrigation is a very suitable sector for donors to support to inject a more practical and less theoretical dimension into their shadow world. The shadow world is generally a better reflection of donor (and recipient) aims but at present it generally lacks credibility. Until engineers can lift the telephone to find the shadow price of cement and be confident that there is financial support for shadow pricing procedures, progress towards economic efficiency will be slow.

Construction

In drafting this section and the following section on operating problems, the writer has had the benefit of various World Bank and United States AID confidential reviews in addition to his own evaluation experience and other published materials. The main non-official documents are listed in the references. In identifying problems, the specific project is not mentioned although the country where evidence accrued may be cited. Research and training in this area must be encouraged because such understanding has to precede action.

Various technical problems are also creating operating difficulties. Drainage is generally lacking and where present it is poorly maintained often endangering the health of both farmer and his soil. Land levelling is often poorly executed.

Restriction on illegal or excessive withdrawals from canals and groundwater are seldom effectively enforced. Collection of management information and communicating it to decision-makers is defective, resulting in operating inefficiencies such as a slow reaction to rain. These examples are illustrative and not in any sense comprehensive.

A more comprehensive review of operating problems is presented in a recent CGIAR study Team Report on Water Management, Research and Training. The Study Team visited eight countries where irrigation is important and confirmed the general view that severe irrigation operating problems exist. The view of various Indian officials on difficulties encountered in their work. To show some details, S. Hashim Ali listed the major 557 defects found in a rapid survey or 232 villages in one command in 1979.

These findings are not greatly dissimilar to findings in other countries. To correct these deficiencies and reform the system is not excessively costly. It is not surprising that the rates of return to rehabilitation projects are so high. There is clearly an important role for aid donor assistance which is only being partly filled at present. The facilities need to be modernised, rehabilitated, extended and drainage provided. This together with multi-disciplinary management training would play an important part in helping farmers achieve the potential productivity that eludes them at present.

Operating phase

Despite widespread acceptance of the complexity of the irrigation development and operation problems, it is human nature to seek simple solutions or an apparent panacea. Currently disappointing performance is widely attributed to "management" difficulties. For example chandler reports that: "The philippines, Taiwan and South Korea all developed similar programmes to increase rice yields, involving new technologies (irrigation, new varieties, fertilizer, pest control), credit, and extension services". He states that "the differences in success between the three countries' rice yields of (less than 2 tons\hectares, 4.5 tons\hectares, 5.9 tons\hectares respectively) appear to stem from differences in irrigation management".

Management is an ambiguous, elusive concept. Each of us probably has a personal understanding or definition. In agricultural economics the contribution of management has often been estimated as the positive unexplained residual which occurs after estimating the returns to land, labour and capital. Management is generally viewed as a positive influence. In developing countries in general, and in irrigation in particular, management is often seen to be a negative factor, a constraint rather than a contribution. This perception is hard to confirm or reject empirically since it is generally agreed that, at least for agricultural projects, one cannot judge management simply by its performance, because of the number of unpredictable natural and economic constraints outside management control.

Disappointing operating performance stems from many sources and not just management. There are shortfalls in finance, poor condition of some infrastructure, faulty liaison with farmers and insufficient institutional support. Studies have revealed sub-optimal performance in area cultivated, intensity of land use, inequitable water distribution, inefficiency of eater use, crop yields and environmental, impact, particularly in public health and soil salinity, and all this cannot be laid at the door of management.

Physical and operational inadequacies in canal systems (As cited by Indian officials) (1)

Mains system Inadequacies

Faulty original design

— The present level of competence in planning and preparing projects is too low.

— The system most often lacks sufficient control structures or control structures are not properly designed.

— Canals are often of inadequate capacity in relation to peak demand.

— Discharge capacity of project outlets often too low in relation to the irrigation stream that can be handled by farmers.

Faulty operation and maintenance

— Canal capacity less than original design capacity owing to poor maintenance.

— System operation often based on water availability rather than water demand.

— Head-end farmers get oversupply of water in initial years and resist later reduction of supply.

Faulty modernisation procedures

— Measuring devices often lacking.

— Phasing often unsuitable, e.g. overlooking that first effort should be to restore system back to original design capacity.

— Techniques for introduction of warabandi (rotational water supply) often not properly thought or implemented.

— Irrigation Department staff lack an adequate manual to guide them in diagnosis as a step in planning system modernisation *On-Fara System Inadequacies*

Faulty institutional arrangements

— Lack of co-ordination among key agencies especially the Irrigation Dept., the Command Area DEvelopment Authority and the Agriculture Dept.

— Failure to appoint a head to direct (rather than merely co-ordinate).

— Lack of follow-through on Central Government

recommendation to create separate Operation and Maintenance Wing in the State Irrigation Depts.

— CADAs have inadequate administrative powers.

Faulty personnel policies

— Executive officers rotate too frequently.

— CADA administrators often too junior or fail to have leadership qualities.

— The CADAs do not provide adequate career opportunities.

Faulty manpower planning

— Lack of facilities for in-service training in water management.

— Lack of adequate curricula on water management in agricultural and engineering universities.

Faulty management practices

— Executive officers overloaded with administrative tasks.

— Inadequate exchange within the country of lessons from experience re operation of water users' co-operatives.

— Lack of clear targets for guidance of managers.

Faulty on-farm practices

Plot-to-plot irrigation as widely practised is inefficient.

Known technologies of efficient water use are not being applied.

The States are slow in responding to Central Government's request to extend field channels at

State expense to 5 to 8 ha. blocks (rather than about 40 ha. as at present).

Faulty drainage and systems

Drainage system often omitted or inadequately designed.

Farm-to-market roads often lacking.

Conjunctive use not sufficiently considered as means of reducing water-logging and salinity while supplying supplementary irrigation water.

Research is lacking (and therefore farmers are not receiving advice) on appropriate farming systems including choice of cropping pattern either under existing conditions (with deficiencies in the water system) or under improved conditions.

Table 2. The main defects found in the quick survey of a canal system in july 1979

	Reasons
(A) Insufficient supplies	High seepage in canals. Conversion to wet in early reaches. Silting. Minor incapable of drawing design discharge. Never received water in ten years. Design discharge not released because of weak bunds.
(B) Defects in outlets	Not fixed. Fixed in wrong place.
	Fixed but defunct.
	Water flowing one foot below vent.
	Unauthorised pipes.
	Broken pipes.
	Open cuts widened by upper farmers.
(C) Insufficient control	Minors having both wet and

	intermittent pipes. No control structures for nearly all minors and many majors. Raising crest of drop necessary. Cistern and regulating arrangements not provided at off take points. Minor breached due to insufficiency of syphon vent way.. Irregular and untimely supplies. Wrong creation of minor offtakes. Insufficient capacity of minor.
(D) Field channels	Not excavated. Insufficient extent not reaching each survey number. Open cuts without structures or drops. No culvert over field channel crossing cart tracks. Constructed by obliterated or damaged. Narrow channel incapable of taking full discharge. rapid flow causing erosion. ramps at crossings not provided Improper alignment. Insufficient drops and distribution boxes.
(E) Lands not getting water	Uncommandable area localised. Damaged drops\cross structures. Full supply level cannot be maintained. Weeds in canals. Desilting not done. Minor widened by farmers for using as cart track. Not a single acre. getting water under one minor. Inadequate supplies in over 50 per cent minors. Badly leading system. Major not excavated up to required bed level, hence design discharge cannot flow. Breaches not repaired. Minors without drops and structures. Minors, pipes silted. Shifting of minor head sluice necessary.
(F) Water logging	Due to heavy seepage from minor. Due to conversion of lands into wet in upper reaches.

Environmental impact

At the risk of breaking long tradition in official documents the following doggerel is commended for study.

An extract from a ballad of ecological awareness

The cost of building dams is always underestimated— There's erosion of the delta that the river has created, There's fertile soil below the dam that's likely to be looted, and the tangled mat of forest that has got to be uprooted.

There's the breaking up of cultures with old haunts and habits loss, There's the education programme that just doesn't come across, And the wasted fruits of progress that are seldom much enjoyed, By expelled subsistence farmers who are urban unemployed.

There's disappointing yield of fish, beyond the first explosion; There's silting up, and drawing down, and watershed erosion. Above the dam the water's lost by sheer evaporation; Below, the river scours, and suffers dangerous alteration.

For engineers, however good, are likely to be guilty of quietly forgetting that a river can be silty, While the irrigation people too are frequently forgetting. That water poured upon the land is likely to be wetting.

Then the water in the lake, and what the lake releases, Is crawling with infected snails and waterborne diseases. There's a hideous locust breeding ground when water level's low, And a million ecological facts we really do not know.

There are benefits, of course, which may be countable, but which have a tendency to fall into the pockets of the rich, while the costs are apt to fall upon the shoulders of the poor. So cost-benefit analysis is nearly always sure, To justify the building of a solid concrete fact, while the Ecological Truth is left behind in the Abstract.

This brilliantly illustrates the sad fact that in the understandable frantic haste for development, losses to the environment, public health and human social\cultural values are almost inevitably incurred. The losses are real costs to society that vary widely in magnitude. They may stem from poor planning, poor prediction or poor management, or it may be that ways of alleviating or minimising their impact are not known. In the planning calculus these facts must be addressed even if it raises complex empirical and methodological issues. If the potential losses are ignored they will not go away and indeed one day must be faced. Delayed action may to more costly in the long run. An early assessment of likely environmental impact may facilitate exploration of alternative designs to avoid, mitigate or minimise undesirable effects.

In irrigation the ecological consequences of proposed changes in land-use and population distribution should be carefully assessed. Such an assessment would require a thorough knowledge of the regional ecosystem and the factors maintaining long-term balance. Assessing the effects on the regional resource base of the proposed construction and irrigation farming may appear obvious but scrutiny of project appraisal reports shows that it receives cursory and often apparently careless treatment. Planners certainly recognise that experience suggests that bringing water to relatively large geographic areas for irrigation projects will require construction camps which will probably become permanent cities, will create what are often unhealthy farming conditions, will disrupt existing wildlife and animal migration routes, will produce a haven for crop pests, will lead to soil salinity or waterlogging, will create new villages with new working patterns for the inhabitants and often new forms of housing and diet.

Unfortunately this recognition of potential problems is seldom properly integrated into the planning process. Environmental considerations in normal planning procedures often fall between the remit of various technical specialists and are not tackled until damage is severe. Environmental problems tend at best to be cured rather than prevented.s In part this is because we have limited vision of the environmental impact. There is a shortage of data and, given the complexity of the ecological linkages, to obtain the necessary data is costly. It is also difficult to translate some of the ecological concepts into the economic rationale of planning—to obtain some notion of cost and benefit and then, given the high degree of risk and uncertainty associated with environmental predictions, to derive decision rules to guide judgment and action.

Irrigation planners and managers have to recognise the very diverse impact of their technology upon the environment. They need to consider the effects more broadly than hitherto and develop and apply more comprehensive qualitative criteria than is traditionally considered in irrigation project appraisal.

Health

Health is a special instance where environmental impact can be neglected. In the enthusiastic drive for direct agricultural benefits the health problems risk being discounted. Dams, canals, watercourses and drains all present health hazards. Total prevention of disease may be an unrealistic goal but thoughtful study and planning can reduce disease problems at bearable costs. Medical, particularly epidemiological, advice should be taken. The unrealistic purist medical, advice which would counsel against almost any development which increases health risk has to be avoided. For example it is known that almost without exception irrigation in an area where schistosomiasis occurs increases the endemicity. If the irrigation is perennial the increase in prevalence can be very high. Nevertheless, projects will go ahead because potential economic benefits are high. However, in these circumstances general preventative measures are both a moral and economic imperative.

Sensible preventative measures are well-known. Selection of the construction site and provision of services of water and sewerage for staff and labourers can be improved by planning. Labourers are often drawn to irrigation sites from distant areas and if from hilly areas they may lack immunity to malaria or at least to local strains. Migrants may bring diseases from their home area. Irrigation projects are a good focus for primary health care, for basic public health measures such as table water, sanitation, vector control around settlements, health education, attention to nutrition (particularly on monocrop projects).

Special attention needs to be given to control of the invertebrates that live or breed in water and which transmit the diseases of schistosomiasis (spread by snails), onchocerciasis (simulium flies), malaria, arbovirus fevers and filariasis (mosquitos). Planners need to appreciate risks, understand transmission mechanisms, minimise the contact between man and unprotected water, design to prevent man-made breeding grounds arising, provide reasonable finance for preventative measures in original designs and, most important, maintain water facilities, particularly drains, so that standing water does not exist—at least not close to settlements.

Other neglected areas

Farm management, water management gap

Irrigation engineers study primarily construction and focus mainly on dams and canals. Agriculturists study crop science and tend to specialise in single crops or specialisms such as crop protection or soil science. Farming systems have been discovered and developed only recently as an area of special study, largelyl by farm management economists. Field engineering—culverts, gates, ditch construction, land levelling, field drains and the like-has been neglected. Often it is left to farmers who may lack the finance, knowledge or collective will to completely execute these necessary works. Management of the field level facilities to serve the complex and varied need of multiple cropping farming systems is at present the province of on specialist although it is clearly vital that water supply and on-farm water use are well-matched. A study of 91 irrigation projects in 1978 revealed efficiencies of the diverted water at around 30 per cent only. Improved watercourse and field management would increase this readily by 50 per cent.

Illicit payments

Even a rapid appraisal of farmers on any irrigation scheme is likely to reveal that there is an unofficial system of illicit payments for services including:

— bribes for additional water or ignoring illegal crops;

— extortion for delivery of allocated water—a kind of insurance or protection racket.

More extensive informal discussion with engineers and farmers is likely to reveal unsubstantiated evidence that it is possible to earn "commission" from new schemes contracts or from O&M contracts. Such transactions have been long established practice. Whitcombe quotes a 19th century private investigator who provided:

— "documentary confirmation of the oral statements of various lambardars (village officials) that illegal canal imposts have been extorted ever since the introduction of canal-irrigation, or that they have paid the same,or known them to be paid, ever since they were lamberdars, and that their fathers paid before them.

— In fact there is not a Canal Deputy Magistrate who did not share in the plunder of the people when ziladar (an official heading the local canal administration), and not a ziladar who is not concerned in extortion."

Robert Wade has provided the most recent, most comprehensive account of the scale, workings and impact on irrigation efficiency that has been published. It is based on extensive fieldwork in south India. In an extremely thorough review he describes how O&M contracts are tapped for a share of contractors' profits; for allowing shoddy, sub-standard work (e.g. removal of less silt than in a canal clearing contract); by subversion of tendering procedures; by advancing credit to contractors; by diverting illegal earnings to business partnerships; by obtaining privileged access to rationed items such as cement; by selling base assured water supply and extra amount at charges according to season, drought conditions, whether zoned or unauthorised land is used, profitability of crop, location of land, and political power of village. He discusses methods of payment in cash and kind, and who gets what share and who actually handles money. The exchange of money through personnel transfer and the way in which money gets into the political system is outlined.

This writer has field experience in several countries on canal irrigation schemes, that tends to confirm that the process Wade describes is, with variants, found elsewhere. Given that this process creates uncertainty and unreliability it must affect farming investment. Indeed, as Wade points out, to maximise bribes, engineers have to persist with ad hoc unannounced cut-offs, water rotations and create uncertainty. He points out that operating engineers may blunt reform if it weakens their financial interests. Running the complex water financial market must detract from their professional activity and tapping the O & M budget must produce sub-standard work. If it is true that part of the revenue percolates to politicians, the fate of reforms in the legislature (such as increasing official water rates) is not likely to be as vigorously promoted as is desirable.

The main reason for this somewhat lengthy discussion is that in very few of the official donor evaluation reports is the subject mentioned and in none is it given any prominence. But it probably does constitute a major impediment to efficiency. Bribery is not a lubricant to the operating system, it is corrosive. In an era where the quality of aid is of increasing concern donors should face the vital questions this research raises. Donors could perhaps play a role in the two most attractive solutions advocated: first, to promote irrigation user associations as a countervailing voice; second, to encourage professional training and a cadre of people dedicated to professional norms of efficient canal operation. Such small endeavours could have useful impact.

Operation and maintenance

Previous sections set out some of the causes of neglect. The CGIAR study Team suggested that without improved O & M, yields of crops such as paddy rice will not rise from 1.5-3.0 ton\hectare to the 5-6 ton\hectare which is already feasible on a project scale. Efficiency of water use is unlikely to rise above 30 per cent without new efforts. They cite Pakistan data to show that the cost per unit of water saved through watercourse improvement was only one quarter of the cost of new supplies. Improved O & M generally improves equity as those that suffer most from deficiency are the poorest. In the current aid climate, donors will hesitate to support projects with inefficient operations.

Corroboration that water management is neglected comes from a lengthy 1981 analysis of 30 World Bank-supported irrigation project systems in 15 countries. This report concluded: "Overall, water management in the set of irrigation projects under review was found to have received inadequate attention. Insufficient provisions for the systems' operations and maintenance were made at appraisal; insufficient action taken during implementation. Analysis of water management issues in completion and audit reports—as well as in the appraisal reports-tended to be incomplete or superficial; quantitative data was sparse and fragmentary. On the basis of the available evidence, water management was found to be good in only 9 of the 20 projects for which information is available. Some shortcomings and considerable room for improvement exist in 7 other cases; definitely poor water management affected 4 projects. Water supplies proved inadequate in 10 cases, and unreliable in 6; water distribution, inequitable in 3; water losses, excessive in 5." The report concluded that there has to be simultaneous design of O&M system with physical works. This should include operating manuals and staff training. Experienced well-established national agencies perform better than new project authorities. Extension agencies perform better than new project authorities. Extension work at the field level tends to be a weak link and needs special attention. The project authority has to provide tertiary canals and below and to check on construction and operating quality. Maintenance workshops, equipment and machinery have to be provided and, most important, financial arrangements to sustain them have to be assured. The financial mechanism has to have a fall-back position in case the intended source of finance fails. This arrangement must also have protection against inflation. Indeed the Bank stresses that finance for O & M has to be provided and water charges, land taxes, sales or value-added taxes substituted as and where the selected mechanism does not meet requirements. Donors should learn the lessons of this extensive experience and include satisfactory donor or recipient provision for operation and maintenance—not just unenforceable covenants or vague promises.

Links with rural water supply

The United Nations Water Supply and Sanitation Decade is nearing the end of its second year. The main aim is to provide a safe drinking water supply to all by the year 2000. Field research in Kenya conducted by the writer in the early 1970s suggested that health and economic benefits from improved water supply were hard to sustain even when household connections were provided. This was because of a paucity of complementary facilities such as health education, agricultural extension, credit and marketing facilities and defective O & M. One area was found where real benefits occurred providing the scheme delivered water at least intermittently. When additional water goes into a farm the vegetable production in the kitchen garden increases. This minor irrigation provides protective foods for the family and often a cash income from market sales for the female farmer.

We have already described how irrigation can create health hazards. However, it can also create opportunities for benefits. Research into rural water supplies has shown the importance of access to large quantities of water for controlling a set of water-related disease, notably the diarrhoea, eye and skin infections. In irrigation schemes it is possible to design public access to high quality and large quantities of water and achieve important health benefits.

There have been insufficient links between the water Supply and Sanitation Decade and Irrigation Development. Irrigation presents health hazards but it can also provide the economic base and access to water that can enable the Decade goals to be realised.

The long-term

Politicians often have very short time horizons. In the extent that this is true they will favour policies and projects with rapid impact. They will neglect desirable projects such as forestry, rehabilitation or replanting of perennial crops, soil erosion control and other conservation measures. In the irrigation field some preventive drainage projects, land reclamation projects and public health maintenance measures would fail to pass political tests because of the long-term nature of their benefits. Future costs, such as high operating expenditure or expensive replacements, will tend to be ignored in such a political climate. Such a mentality, giving much the greatest weight to current benefits, is given theoretical legitimacy by current social cost-benefits analysis procedures, which, guided by some often spurious notion of opportunity cost, favours high rates of discount. An alternative rationale that future generations will have higher income, hence lower marginal utility, and therefore that they should receive lower weighting can be regarded as equally dubious on moral and empirical grounds. The moral objections to interpersonal utility comparisons can be extended to inter-generational grounds. However more pertinent is the practical question of poverty alleviation. The writer knows of no studies which forecast income growth in the next two to three decades of the poorest groups in low income countries that can take them out of the range where they will have equal utility from marginal income than the current poor. Hence he cannot defend discounting future benefits of projects for the poor on these grounds. For aid donors he would favour using relatively low rates of discount (in line with actual rates of return, say about 10 per cent) and giving some weight on the final analysis, where hopefully many criteria are applied, to the low rate of return project such as those mentioned which gives large but long-term benefits, greatly in excess of costs.

Investment policy choices

In this section issues are identified which are considered to be of direct relevance to aid donors. In order to stimulate discussion the issues are polarised and posed as alternatives. It is accepted that sometimes specification in this way is somewhat contrived, artificial or inaccurate, but it helps focus attention on a sub-system and it can thus assist in forming an agenda for discussion. The alternatives posed are sometimes linked. For example private development leads to considering small-scale projects-if drainage is considered this is large-scale, public sector.

It is important not to be too dogmatic about appropriate irrigation investment, management and other policy matters. The Malaysian experience confounds much conventional wisdom. For example, higher levels of investment in density of distribution channels is commonly held to lead to be associated with more equitable distribution of water and higher yields. This does not hold in Malaysia. Furthermore, even investment in concrete lining of the field channels does not reduce O&M expenditures. Nor do farmers at the tailend, distant from the source, have a less satisfactory supply of water than those at the head. Most surprising there is no clear tendency for smaller paddy farms to have higher yields nor for tenants to have lower yields than owner-operated farms. These results stem from carefully executed field studies throughout Malaysia and the results are quoted to show the necessity for thorough research to monitor and test preconceived notions before large-scale public investments are undertaken.

New projects or modernisation and rehabilitation

The most important expenditure is to keep existing projects in full working order. The next choice is between new projects and rehabilitation. New projects are becoming increasingly expensive to execute and suitable sites are more limited. The remaining soils are less fertile. On the other hand, for the existing schemes, after years of neglect of maintenance, some form of rehabilitation is generally necessary. Deferred expenditures on such matters as field drains or watercourse control structures are now overdue. Operation difficulties are often due to old or deficient structures and communications. The main capital expenditure is a bygone or sunk cost giving high rates of return to modernisation and rehabilitation. Rehabilitation and modernisation should have priorities over new projects. Aid donors should reconsider their attitude towards local costs and recurrent finance for rehabilitation projects.

Drainage or more irrigation

Various forms of damage from inadequate drainage are still all too often regarded as an unexpected indirect cost of irrigation development. For example, in Egypt, with its long experience of irrigation, there was disappointment verging on surprise at the extent and form of the deleterious effects of the large additions of irrigation water from the Aswan dam, first on the groundwater regime, then later on crop yields, The damage has forced the Government to adopt a nationwide drainage programme that has absorbed the major part of the Ministry of Irrigation's capital budget in recent years.

In most irrigation areas, drainage and reclamation projects are needed now. FAO estimates that 50 per cent of the world's irrigated land is salinised to the extent of affecting productivity. In Iran, Egypt and Pakistan more than 70 per cent of the farmland is so affected. India has 12 million acres affected. Drainage is not undertaken because the effect of water-logging and salinity is generally slow to become apparent; remedial measures are expensive; in already irrigated areas the loss of land and disruption to existing farm structure, roads and canals causes local opposition; maintenance of drains is costly and requires conscientious management.

The best technical means, the operating system, and the economics of drainage are not well tested. Donors could assist with finance, technical assistance and pilot projects. Once this phase is over there will be a major role for aid donors with large resources, a long term perspective and an environmental consciousness. The economics of drainage shares the problems of conservation, soil erosion and three planting. The rates of return are likely to appear low but the instinct is to proceed in spite of this. As was previously noted on long-term irreversible matters the economic calculus sometimes appears fragile and deficient.

Large or small-scale projects

Dissatisfaction with performance of large-scale schemes has led some supporters of irrigation to advocate very small-scale irrigation schemes. Such suggestions fit well with current fashions in development which insinuate "small is beautiful" and that full participation is essèntial. Unfortunately there is some indication that costs of irrigation for very small-scale schemes are often relatively high and the performance relatively poor. (1) However, the overall effects of scale upon capital and recurrent costs, upon cropping intensity and yields is far from clear and research is required to test the relationships for different technologies, cropping patterns, ecologies and management systems.

Some of the most successful irrigation projects in Asia and Africa have been small-scale private schemes often using groundwater or lift irrigation. They use mainly private initiative but they benefits from public support in the form of credit (including credit for hiring labour), agricultural research and extension advice, training of pump mechanics, surveying and so forth. Small-scale projects are often efficiently operated. Where Government revenue is a very scarce resource the minimal levels of financial support that small schemes command is an important consideration. Technical inefficiency of small pumps, excessive levels of investment become secondary considerations.

For water storage schemes, flood protection schemes, drainage schemes and canal irrigation in large alluvial basins the economies of scale are so great that it is hard to ignore them. However, performance of many of these recently constructed large schemes is so far below predicted levels that, given realistic economic analysis, even the least cost solution is likely to be uneconomic. In planning irrigation the small-scale option must be explicitly considered.

Direct investment or price support

In irrigation construction economies of scale can be a snare and delusion. Even where management have the advantage of a modern irrigation network and long experience this is no guarantee of a productive agriculture. For example, in Egypt despite thousands of years of experience with irrigation and the completion in the 1960s of the High Aswan Dam which achieved spectacular results in matters of flood control, regularisation of river flows and power generation, the statistics of both cultivated area and growth in output indicate production increases much below expectations.

Disappointment may not be due to salinity, unreliable water supplies, poor canal maintenance and such technical issues. Agriculture has to be profitable. There is clearly no point trying to enforce water charges or other taxes if farmers do not have a reasonable income.

Research in the last 20 years has established that for the most part farmers are responsive to high prices. However governments all too often neglect these insights and are for their part most responsive to the demands of the rapidly growing urban population for cheap food. In inflationary periods the temptation to fight inflation by holding basic food prices down can rapidly produce an unprofitable agriculture. The availability of food aid or concessionary sales or credits for food imports can provide the mechanism for maintaining urban food supplies. Over-value exchange rates can make commercial imports look articicially cheap. Once a country adopts a regime of a cheap food policy and creates an unprofitable agriculture it is difficult politically and financially to shift markedly from it. There are a number of countries like Egypt where economic management has lost the balance between a profitable domestic agriculture and reasonable food prices. In due course the burden of food import financing can lead to starving the irrigation and agriculture departments of finance-capital and recurrent, creating an inefficient poorly managed system in time providing a rationale for food imports. Sector reviews of irrigation must include an assessment of agricultural pricing and foreign exchange policies.

Public or private development

There is a long-running debate on whether irrigation should be developed in the private or in the public domain. There is inevitably much to be said in each cause—as is usual with such strongly polarised positions.

Public enterprise has had a distinct role in pioneering large-scale irrigation and groundwater projects and in extensive exploratory programmes to define the very nature of the resources. In this respect the public initiative has been an essential pre-requisite without which the private sector would have been unaware of the potential. Insofar as any irrigation system is equitable, the large groundwater projects where wells discharge into established irrigation canals do ensure reasonable equity and, certainly, some of the benefits reach the poverty groups.

Technological progress in the irrigation and groundwater field, in terms of planning and managing projects, is virtually limited to the public sector. Such problems as optimising the design of a system, investigating and testing new materials, providing flood protection, modelling an aquifer, intergrating groundwater with surface water supplies, operating the aquifer as a storage reservoir and controlling saline intrusion are all matters which can be undertaken by a public authority, but cannot be done in the private sector.

Private development, on the other hand, can score on several accounts. It can mobilise capital, which might otherwise lie dormant, and progress is generally far more rapid than in the public sector. The overwhelming advantage of private ownership of water sources and wells is in the matter of maintenance and control of the water.

At the individual source level, the private operator is much more efficient.

Private farmers generally elect for low-cost technology. Their aims are short-term and they therefore employ cheap materials. Risks are high and there are frequent failures, but the private individual is prepared to take these risks whereas a public authority dare not entertain a possible high failure rate. Thus we find high technology strenuously defended on economic grounds in the public domain and the low technology of the private sector equally strongly defended by the appropriate technology faction.

Uncontrolled private development, even with low technology, can lead to excessive investment per unit of area. In groundwater development, with each farmer wanting to own his own well, there are cases of ten or more wells in one area which could reasonably be irrigated by one.

The paradox is thus that private developers are able to install and operate a system efficiently but cannot possibly begin to consider overall management, either of the total investment or even of the resource itself, particularly if this involves integrating surface with groundwater sources. It seems therefore that some form of overall government control would provide the best compromise, with farmer groups operating and maintaining the final distribution systems and tubewells.

Hydro-power versus irrigation

Generally demand for water and hydro-electricity demands match reasonably well. When it is hot, plants require irrigation and pumps and urban areas, electricity. However, conflicts can occur. For example, in Egypt there is a high level of urban power demand even in winter when canal closure for maintenance is necessary. Nearly 4 billion m3 presently spills from Aswan at this time and directly flows to the sea.

Where canal offtakes are above the dam irrigation consumptive use can lower power capacity. Priority at this time has to go to the most economic use and that is usually urban use. If tubewells and lift irrigation are a significant proportion of the power load, then integrating irrigation power needs with overall power planning becomes important. Tubewells and pumps can become an important element in optimising power operation in that it is ideal for peak load shedding. Conversely, once an important proportion of irrigation capacity is dependent upon electric power, then providing a reliable supply is economic. In India today a reliable power supply to rural areas is probably the cheapest, quickest and most efficient irrigation investment.

Power is only one of the inputs which agriculture needs as a reliable, reasonably priced service. Fertilizer, seeds and so on are needs as in any agriculture but given the higher levels of investment, the opportunity cost of non-availability of reliable power is much higher.

The hardware-management mix

The previous discussion indicates that there is considerable evidence that the balance between new schemes—more hardware and improved operation—better management is not ideal. The CGIAR Study Team concluded that:

"Poor performance is often linked with severe deficiencies in irrigation management. These arise from inadequate planning (with narrow objectives, lack of realism, and neglect of non-irrigation aspects), from defects in design (neglecting requirements for flexible operation, on-farm development, and drainage), and especially from deficiencies in operation and maintenance (including farmers' participation, the distribution of water, and financing maintenance).

Good irrigation management requires a multidisciplinary approach,but specialisation leads away from this. In particular, civil enginering training and professional formation reward and value construction and neglect and undervalue the vital activities of operation.

Actions are under way in most countries to remedy the deficiencies but must be greatly strengthened and accelerated if the essential food and social targets which can be achieved through irrigated agriculture are to be met. Many current deficiencies in irrigation management are linked with institutional problems. Institutional changes are a national concern and responsibility but outside support can encourage and catalyse change in conditions favourable to collaboration and where this is acceptable."

Several ideas have been put forward to develop research and training for irrigation management among which a recent proposal of establishing an International Irrigation Management Institute within the CGIRA system. This initiative has now been accepted and the donor community is invited to support a management institute which is likely to be located in Sri Lanka with field research in many countries. The total costs for the first five years are likely to be less than $20 million-about the cost of 2 000 hectares of new irrigation in Africa, perhaps 4 000 hectares in Asia.

Groundwater and conjunctive use

Carruthers and Stoner set out the key economic aspects

and policies which arise in groundwater development. They indicate the very profitable investments available in groundwater and point out the importance of groundwater development for drainage and the largely untapped benefits to be obtained by planning conjunctive use with surface and groundwater. Groundwater in conjunctive use with surface supplies can increase productivity by increasing total water availability, meeting peak requirements of profitable cropping patterns, allowing flexibility in allocation within canal commands, and extending the season and safeguarding extremely dry season flow. Most important perhaps, increases in water availability from groundwater can be used in projects to reform water distribution and increase equity. Aid donors should regard groundwater development as an extremely valuable area for finance and technical assistance particularly where it can be linked to surface irrigation systems.

4

The Political Economy of Irrigation Development

Irrigation benefits and irrigation surplus

The opening up of irrigation possibilities in a community confers three main benefits which accrue to different groups commanding different factors of production:

(a) increased and more stable flow of income from farming made possible by improved yield, opportunity to choose from wider, more lucrative enterprise/ technology mixes, proofing from rainfall irregularity and inadequacy, etc.;

(b) appreciation in the value of land with access to irrigation, which is nothing but a reflection of (a);

(c) increased and more evenly spread farm labour opportunities and improved real wage rates.

(a) and (b) are primary benefits for the sake of which access to irrigation is sought and which accrue to the owners of land benefiting from irrigation. (c) is in the nature of a 'spill-over' and an essential part of the process of realizing the benefits of irrigation. There is generally, but not always, a direct relationship observed between (a) and (c). Use of chemical fertilizers, which normally accompanies irrigation, accentuates weed growth and entails greater labour demand in weeding. Irrigation promotes transplanting in crops like paddy; transplanting absorbs more labour. Improved yields imply more labour required for harvesting, threshing, etc. Irrigation leads to higher cropping intensities resulting in greater labour demand in rabi and zaid, thereby reducing seasonal unemployment and poverty.

The first two benefits broadly constitute the 'irrigation surplus' which, in specific terms, may be defined as the increase in the flow of net farm income that the use of water in crop husbandry generates. Defined thus, 'gross irrigation surplus' is nothing but the total increase in the value added in farming through a switch from rained to irrigated farming; in a static sense, it reflects the shift in crop production function. The value of gross irrigation surplus generated by a new irrigation source in a given setting would depend upon the quality of irrigation service assured, alternatives to the irrigation source under question, and the new crop/enterprise opportunities that it opens up. It thus implies the upper limit to the amount that a water user would, in theory, be willing to pay as irrigation charges.

'Net irrigation surplus' represents the gross value added by irrigation less the nominal cost of irrigation. In most surface water sources, nominal costs of irrigation are vastly different from opportunity cost—either in the sense of the real cost of providing irrigation or in the sense of its true productive value. However, the fact remains that the pricing of the irrigation service is a major device to regulate the distribution of gross irrigation surplus between users and suppliers of water. The farmers' economic behaviour with respect to water use is typically governed by the relative scarcity of land versus scarcity of water. In head reaches of canal commands, for example, where water is plentiful, land is typically seen as the binding constraint on improving family incomes and its productivity is sought to be maximized with liberal use of water. In contrast, where water is viewed as a scarce resource with high opportunity costs at the margin, the 'maximand' in the farmers' objective function is net irrigation surplus per unit of water.

Lifting technology and irrigation surplus

The search for ways to appropriate in irrigation surplus as economic rent' is thus central to the modern history of Indian agricultural development, and in case of groundwater it has run concurrent to the propagation of the green revolution technology. Groundwater irrigation had but a minor role in the political economy of Indian agricultural development until the 1960s; for the 'surplus' generated by groundwater irrigation was relatively small. Traditional Water Extraction devices using human and/or animal energy could lift limited amounts of water per unit of time; a traditional 'rahat' or 'mhot' took between 24 and 50 hours to wet an acre once. At best, thus, the traditional WE technology was suited to merely protective irrigation aimed at supporting the survival of plants in periods of moisture stress in case of delayed monsoons, and occasionally used for rabi wheat, pulse and some cash crop varieties thriving on moisture retained by the soil. Production irrigation was not even needed since most traditional food and cash crop varieties in use were not bread for strong response to increased water application. As a result, most of the older canal systems—for example, in the Gangetic plains—were designed for extensive protective irrigation of large command areas rather than intensive production irrigation of limited command areas. The main purpose for which irrigation was sought lay in minimizing the output risk due to moisture stress, and not in maximizing the 'irrigation surplus.

The more-or-less simultaneous rise of modern Water Extraction (WE) technology and the modern HYV-fertilizer technology of crop production (CP) rapidly altered the priorities of water users. An important requirement of the modern CP technology was timely application of often large quantities of water at crucial stages of plant growth. The traditional WE technology could have met his requirement only on tiny areas due to the stringent limits on the amount of water that could be lifted per unit of time. The modern WE technology with its substantially higher water output per unit of time due to the use of diesel or electrically powered pumps fitted the bill much better. It has been argued, for example by Dhawan, that the spread of the green revolution technology in the northern plains was fulled by the tubewell revolution. The two 'revolutions' have, however, complemented each other; it can be argued equally well that the tubewell revolution was spurred by the opening up of the green revolution technology. Tubewells were not unknown to the farmers in the Gangetic plains in the pre-green revolution era—as a matter of fact, the U.P. state tubewell programme started far back in the 1930s. Diesel engines have been used for lifting water since the early decades of this century in several parts of the country. However, the propagation of modern WE technology speeded up only after 1965 and its geographic spread closely followed the spread of the modern CP technology.

Be that as it may, the key issue is that the modern WE technology and modern CP technology were strongly complementary to each other and together resulted in the generation of massive 'irrigation surpluses' in those areas of the country where they made deep inroads. The key difference between the economics of traditional and modern WE technologies is that the former can lift small quantities of water at a relatively low real cost. To illustrate the difference, the results of a comparative study of bullock bailers' with diesel and electric pumpsets in Pasumpon Muthuramalingam district of Tamil Nadu are presented for irrigating half an acre of paddy, the bullock bailer is the cheapest; but beyond that modern WEMs using diesel engines and electric motors rapidly become cheaper. Indeed, with over 1.5 ha of paddy land, irrigation with the 'bullock bailer' becomes not only prohibitively costly but well nigh impossible since 1.5-2.0 ha.m of water would have to be lifted over a short period of 60-80 days implying 15.20 hours of work per day. With a 7.5 hp electric pumpset, in contrast, the average cost per hour will be quite high at Rs 7-13 if used to irrigate only 1 ha of paddy; but will decrease to Rs 4.07 for 2 ha of paddy and to just Rs 2.38 when used to lift 4.44 ha.m of water to irrigate 4.4 ha of paddy.

Numerous field studies have likewise confirmed the substantially stronger response of HYV yields to water application and fertilizer use than the traditional crop varieties. While it is unnecessary to survey all these studies here, we briefly discuss two which focus on output/income responses to groundwater application. The study by Moorti is set in water abundant northern India; the other one by Deshpande et al. is drawn from water scarce Purandhar taluka of Maharashtra. With respect to the first, needless to say the shift in the crop production function is just one element of the output augmenting effect of irrigation; and the illustration based on Moorti's analysis does not include the additional beneficial effect of the change in cropping pattern towards more lucrative crops and of increased land use intensity. Deshpande's approach, markedly different from Moorti's consider the livelihood impact of a new irrigation source in terms of the increase in the gross value of output per user family.

In water scarce (ws) arid areas, where farming is entirely rainfed, rainfall precipitation is low and erratic, and soils retain little moisture beyond the Kharif season, the overall irrigation impact on the value of output and incomes can be dramatic. In the drought prone Purandhar taluka of Maharashtra, the lift irrigation cooperatives established with the help of Gram Gaurav Pratishthan led to a 2-15 fold increase in the value of annual output per member.

Quality of irrigation service and irrigation surplus

The size of the gross irrigation surplus generated depends crucially on 'quality. The quality of irrigation service includes several factors, but three main attributes of quality in this context are: timeliness, adequacy and reliability. In contrast to traditional farming technology, the HYV seed-fertilizer technology produces best results under a tight, time-bound crop management regime. Deviations from the optimal management regime can result in considerable shortfalls in crop yields.

Effects of lift irrigation cooperatives on the average value of crop output pre member: Purandhar taluka, Maharashtra

Name of the crop.	No. of members	Year of establish-ment	Value of crops per member in year pre ceding the start-up of the project	in 1984	Percentage increase
Mhasoba	26	1980	279	2927	1049
Babawadi	38	1980	212	3482	1642
Shivshankar	26	1982	4404	7958	181
Bhairavanth	17	1982	224	450	200
Phulc	16	1982	412	1881	457

Watering the crops at certain crucial stages of plant growth is important to maximize returns from modern farming. Ideally, therefore, a farmer would want a high degree of control over when to water and how much to water; while planning his crop pattern, he would like to be certain that water would be available at the time at and in the quantities which may be needed at different stages of the crop growth.

Different sources of irrigation vary widely in the quality of irrigation service they can provide and, consequently, in the size of the 'irrigation surplus' that they generate. Studies comparing the performance of irrigated farming under various sources such as canals, tanks, public tubewells, private tubewells, etc. indicate that as the quality of irrigation service improves, land use intensities increase, fertilizer use becomes more intensive, cropping patterns shift in favour of more lucrative crops. All these result in increases in 'gross irrigation surplus.

Tanks tend to offer among the poorest quality

irrigation services on all three counts of adequacy, reliability and timeliness —'supplementary' tanks more so than 'primary' and 'system' tanks. They typically depend upon gravity flow so that controlled water application becomes well-nigh impossible. Most tanks dry up after the main monsoon season so that they cannot be relied upon for rabi or summer crops; as a result, they do little to help improve land use intensity. The most serious problem with tanks is management; in most states tanks are managed by village panchayats as a common property. As with other such common property resources, tanks everywhere have fallen into disuse and disrepair; increasingly siltation in tank beds reduces their capacity to store rain water run off; in many situations, panchayats find it difficult to enforce even a modicum of discipline on users, for instance, in the choice of crops to be irrigated with tank water, or in the collection of dues, or even in closing the sluice on rainly days which could increase the command areas of many tanks by as much as 20 per cent. Although tanks are slowly re-emerging as important in harvesting rainfall precipitation in WS areas, as of now there is little doubt that, by and large, users of tank irrigation in general are only marginally better-off than farmers depending solely on rainfall.

Canals were long thought to be much superior in the quality of irrigation service they can provide. Experience in recent decades has, however, produced much disillusionment about the performance of canals. The only group of irrigators in canal commands who usually have no or few complaint about the quality of irrigation service are in the mid-reach areas of the system. Those at the head-reach suffer from too much water, especially below ground surface, contributing over a period of time to waterlogging, soil salinity and reduced land productivity; while tail-enders of canal systems are in much the same situation as tank irrigators—on most canal systems, the irrigation service, if any, that tail-enders receive is usually neither timely, nor reliable, nor adequate.

In comparison to tanks and canals, extensive evidence suggest that wells offer better quality irrigation service and therefore help generate a larger irrigation surplus. Once problem that all canals have and which wells effectively solve is that of control: in the case of canals, farmers have to use water when the canal managers choose to release it, or when their turn comes, which may be after a week or two or more. Since the moisture requirements of plants have their own logic, distinct from the logic governing the canal system operation, even after paying for canal water, farmers prefer to make supplementary use of well water when possible. Though often much more expensive, access to well water offers farmers the much needed control over the timing and quantum of water application that they do not enjoy with canal irrigation.

In an important study which highlights the differential out-put impact of tank, canal and well irrigation, Dhawan analysed survey data from four Indian states: Punjab, Haryana, Andhra Pradesh and Tamil Nadu. His conclusions summarized in table indicate that the difference in land productivity under well irrigation and that with rainfed farming is over twice the difference observed between canal irrigation and rainfed farming; and tank irrigation adds the least to land productivity under rainfed cultivation.

Output impact of tanks, canals and groundwater: tonnes of foodgrain per net irrigated ha additional to the rainfed yield

	Groundwater	Canals	Tanks
Punjab	4.4	2.1	-
Haryana	5.3	2.0	-
Andhra Pradesh	5.2	2.9	1.5
Tamil Nadu	6.0	2.1	1.8

Clearly these differences in land productivity in Dhawan's analysis conceal the differences in land use intensities, in the use of fertilizer and other inputs and in cropping patterns amongst farms dependent on different irrigation sources. The analysis by Rao of a region in Karnataka in the mid-seventies showed that fertilizer use as well as land use intensities respond strongly as the quality of irrigation service improves. In Rao's analysis too, among tanks, canals and wells, wells produce highest land use intensity; also, farmers using well irrigation made more intensive use of fertilizer than any other group except the one which had access to perennial canal irrigation sources.

Where they have access to canal as well as well water, farmers in many areas conjuctively use canal water for its low cost and

Quality of irrigation service and the level of land use intensity and fertilizer use: Southern Maidan Karnataka

	Land Use intensity (%)	Fertilizer use (Rs/acre)
Tank	85	66
Canal irrigation I	102	247
Canal irrigation II	123	132
Canal irrigation III	147	408
Well irrigation	197	304

well water for the control it offers. In the command area of the Mahi Right Bank Canal in Kheda district of Gujarat, Kolavalli found that such conjective use of canal and well water enabled irrigators to substantially improve land use intensity. As we noted earlier, farmers located in the middle reaches are better-off than those either at the head or the tail and manage highest land use intensity with or without well water.

Land use intensity under canal irrigation and conjective water use: Mahi Right Bank Canal Command, Gujarat

Location of sample farmers	Canal irrigated farms	Farms using canal and well irrigation
All farms	1.84	2.09
Head	1.83	2.02
Middle	1.99	2.23
Tail	1.75	2.00

Kolavalli's analysis also suggests that better quality of irrigation service enjoyed by 'conjective-use-farmers' was directly transformed into substantially greater irrigation surplus by all but the marginal farmers.

Gross margins from canal irrigation and conjective water use: Kheda district, Gujarat

Farm size (ha)	Canal irrigation	Canal and well irrigation	Difference
0-1	4253	3807	-446
1-2	3743	4801	1058
2-3	1648	3873	2225
34	1951	5586	3635

In well irrigation too, several alternative arrangements exist. Those who have the resources invest in their won WEM to gain direct control over the irrigation service. Complete control is, however, very difficult in areas where landholdings are fragmented. A farmer may put up a WEM in one or at best two of the larger parcels; but these may not be able to irrigate his other parcels located in other parts of the village. A farmer may use a state tubewell to irrigate his parcels which fall within its command. Or, more frequently, he may buy water from neighbouring private WEM owners. In all these cases, the quality of irrigation service offered may vary. A study conducted in Pakistan in the late 1970s compared yields obtained by farmers dependent on rainfed farming, on state tubewell water supplies, on supplies from private WEM owners and on their own WEMs. The study hypothesized, as we have done, that the 'quality' of irrigation service determines irrigation productivity and found that farmers' own WEMs followed by private WEM water supplies offered better 'control' to irrigators than state tubewells. As with Kolavalli's analysis of conjective water use in Gujarat, here too, superior 'quality' of irrigation service was instantly translated into higher productivity and, presumably, larger 'irrigation surplus.

Control over tubewell water supply and crop yields: Pakistan (1978)

Water supply situations	Averaged yield			
	wheat		rice	
	No. of farmers	kg/ha	No. of farmers	Kg.ha
1. No control (rainfed)	170	1681	75	1308
2. Fair control (public tubewell)	33	1868	13	1775
3. Good control (purchased supplies from private tubewells)	133	1962	35	1962
4. Very good control (own tubewell supplies)	42	2242	9	2148

State tubewells in most parts of India are no different from state tubewells in Pakistan. As a class, state tubewells everywhere are afflicted by the same problems of the inefficiency, poor maintenance and long shutdown periods, erratic and insufficient power supplies, the dominance by the powerful, right-of-way problems, arbitrary behaviour of the tubewell, operator and the strong propensity of the system towards corruption. As a result the quality of irrigation service offered by state tubewells ranges from indifferent to poor except for those few who can use them as their own private tubewells. Understandably, where feasible, farmers located in the command areas of state tubewells have increasingly began to use water purchased from nearby private WEM owners even though it may be several times more expensive than the heavily subsidized state tubewell water. Evidence suggesting such preference has been recorded by several researchers: for Uttar Pradesh by Mellor and Moorti and Pant, for Bihar by Sharma, for Gujarat by Shah and Asopa and Dholakia, for Tamil Nadu by Copestake.

The key difference between state tubewells and private tubewells is of management and accountability to clients. A state tubewell operator is in reality accountable to no one, for he can neither be punished nor rewarded by the community he is meant to serve; a long and avoidable shut-down period imposes no cost on him. In contrast, for a private WEM owner, selling water is often an important commercial operation. His superior performance results from the incentive to stimulate the demand for water among his neighbours and to maximize the utilization of his WEM.

It is therefore not surprising to find much evidence suggesting that the irrigation service provided by private

WEM owners is usually of a high quality. Pathek et al. for example, found irrigation with private tubewell water supplies distinctly more productive than state tubewell supplies. There is at least one study based on a large sample survey carried out by the NCAER in three Indian states during the mid-1970s which suggested that the productivity impact of purchased private tubewell irrigation may in fact be higher than the use of own tubewell water supplies. The main results, that the farms using purchased groundwater obtained higher yields/ha in all the three states. The difference was as high as 88 per cent in Gujarat.

The figure 2, which is purely illustrative, pieces together the empirical evidence reviewed so far, and attempts a unified picture of the relationship between the value of gross output, various cash and imputed costs and the 'irrigation surplus' generated per hectare of land under five different irrigation sources. The gross value/ha curve is down as a segment for each irrigation source to stress that while the best of a superior irrigation source tends to be better than the best of a source inferior to it, its worst may not be better than the best of the inferior source.

Differential impact of own versus purchased well water on fertilizer use and crop yields: Andhra Pradesh, Gujarat, Uttar Pradesh (mid 1970s)

	Andhra pradesh		Gujarat		Uttar Pradesh	
	Well Owners	Water buyers	Well owners	Water buyers	Well owners	Water buyers
Crop yield:						
Rice equivalent (kg/ha)	2429	2649	1494	2815	1564	1627
Number of Waterings	5.4	4.42	4.35	2.11	2.50	2.40
Fertilizer use	107	119	54	69	44	27

To illustrate, farmers in an arid region and using a highly unreliable tank irrigation source may produce less per ha than rainfed farmers in a good rainfall region. Likewise, good, well managed tanks may be better than tail-ends of most canals; and mid-reach farmers in many canal commands may produce more than farmers dependent on state tubewells.

As the quality of irrigation service improves, farming tends to become more intensive in the use of all inputs such as fertilizers, pesticides and labour; however, the increase in the costs of these inputs rises less rapidly than does the value of output. Land rents also tends to rise as irrigation becomes available. The shaded area abcd represents gross irrigation surplus which rises steeply with the improvement in the quality of irrigation service. Much of the politics of irrigation concerns the generation, manipulation and distribution of this surplus.

Distribution of irrigation surplus

In theory, if a market existed for irrigation services, a user would be willing to pay anything up to the 'gross irrigation 'surplus' that he could generate with the quality of irrigation service. being offered. Such markets, however, do not exist in case of all irrigation sources. The question of recovering and applying a portion of the irrigation surplus' to defray the cost of creating and maintaining irrigation sources such as tanks and canals is settled essentially by political processes. A good deal of social research in irrigation has shown that these processes are seldom efficient. Although maintaining an irrigation system may require only a small fraction of the potential 'irrigation surpluses' that its users generate with its help, the experience so far suggests that public (and most communally managed) irrigation systems—canals, state tubewells, and even tanks—are able to recover from water users less than even the Q&M costs. As a result, in publicly managed irrigation sources, the bulk of the irrigation surplus' generated is captured by the users of water and is distributed in proportion to land owned by them in the command.

In flow irrigation systems, difficulties in volumetric measurement ad exclusion pose peculiar technological problems in cost recovery. However, in state tubewells measurement and exclusion are easier than in case of canal and tank commands. Yet, STWs in most states combine low water tariffs with an different quality of irrigation service. This is explained in part by political 'imperatives', but more importantly by the fact that in most areas STWs have had to compete with private WEMs and cutting subsidies may make it virtually impossible for them to remain in business.

In contrast, the distribution of gross irrigation surpluses generated by private tubewells is determined almost entirely by market procsses. Since the cost of installing and operating a WEM is completely internal to a WEM owner, the required proportion of 'irrigation surplus' on his own farm automatic ally gets applied to covering the cost of the irrigation service. However, the nature of the WEM economics is such that average pumping costs steeply decline as utilization approaches capacity. Further, since WEM capacities vary only discretely, only those who have holding sizes large enough to match the WEM capacity can maximize their 'net irrigation surplus. Others have to rely on selling water to neighbours in order to benefit from a reduced average cost.

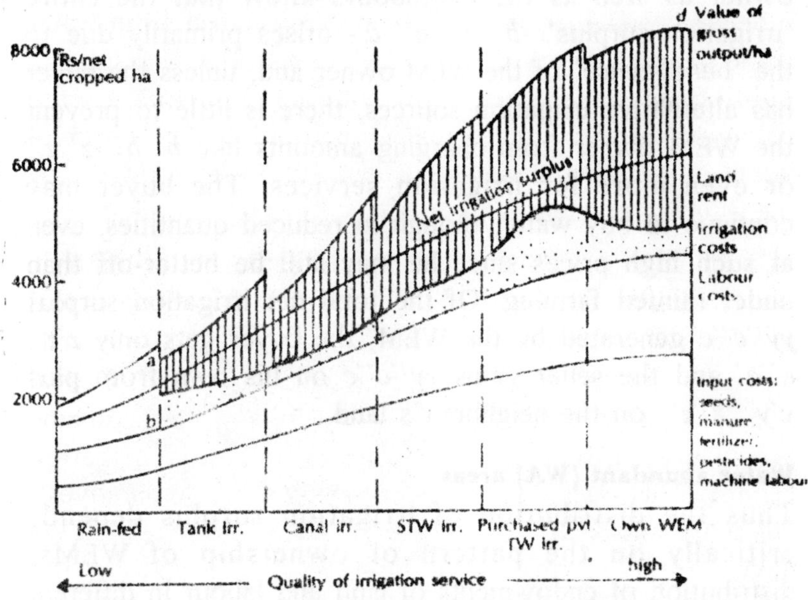

Figure 2. Quality of irrigation service and size of the irrigation surplus

Reducing average cost—or, sharing the overheads of operating a WEM with neighbours—itself could be an important enough business goal for a WEM owner. If a WEM owner irrigates only his own farm, then the gross irrigation surplus that he generates is shown by the area bb′c′c: the average cost of irrigation is *y′x′* and total irrigation cost. *yy′x′x*. His net irrigation surplus is *xx′ c′c*. If he chooses to irrigate an additional *a a* ha of his neighbours' fields, his average cost declines along *x′y″*. Even if he charges his neighbours a price equal to the reduced average cost of *b″ y″* that is, a total of *b′ b″y″ y′*, he stands to gain from increased net irrigation surplus on his own land by *yy′ x′ x′*. His neighbours and, of course, society as a whole, benefit by *y′ y′c″c′*

There are, however, situations where both the WEM owner as well as the neighbours know that the entire 'irrigation surplus' -$b' b'' c'' c'$- arises primarily due to the 'beneficence' of the WEM owner and, unless the buyer has alternative irrigation sources, there is little to prevent the WEM owner from charging amounts like $b' b'' z' z''$ or even more for irrigation services. The buyer may continue to buy water, though in reduced quantities, even at such high prices since he may still be better-off than under rainfed farming. Of the total net irrigation surplus $yy''c''c$ generated by the WEM, the buyer gets only $z'z'' c''c'$ and the seller gains $yy''c''c$ on his own from plus $y'y'' z''z'$ on the neighbour's land.

Water abundant (WA) areas

Thus the distribution of irrigation surplus depends critically on the pattern of ownership of WEMs, distribution of endowments of land and labour in different families, and the structure and performance of, and the terms of transactions prevailing in, fragmented water markets. From the distributional angel, three distinct flows of income are important: net irrigation surplus comprising the difference between gross irrigation surplus and the price charged for irrigation; water sale surplus or rent comprising the difference between price charged for water and the full economic cost of water production; and the wage income surplus consisting of increased wage payments per ha of irrigated over unirrigated land.

We can formulate a priori several hypotheses about the distributional outcomes under different conditions, we use different signs to indicate the direction of change in various components of irrigation surplus under autarky, under monopolistic water markets, under competitive water markets and under a communal monopoly. It may be noted however, that the number of '+' 'o' or '*' signs suggests the direction or ranking of different states far more confidently than the extent of quantitative variations across states.

Consider two extreme situations; first, where private WEMs are owned by large farmers who, for some reason, choose not to sell water to their neighbours although enough pumping capacity exists to irrigate all of a village's farm land. All the three income flows NIS, WSS and WIS assume low values in relation to their potential. NIS is low because WEMs are underutilized; only a part of the irrigable land is irrigated; and average pumping costs are high. If only small holders own the WEMs, autarky would produce only worse productivity results; for area irrigated will o be even smaller and average pumping costs higher; WIs will assume a lower value and wage earners will be worse-off than when WEMs were owned by large farmers. WSS in either case is zero.

At the other extreme is the second situation—of competitive water markets-where WEM owners freely sell water to neighbours at a price high enough to cover their full economic cost of pumping. In this case, too, WSS becomes zero since the water market is competitive. But NIS and WIS assume their maximum possible values. All of the village's farm lands are irrigated; WEM owners and buyers both face the same water price schedule; the NIS generated is further increased because installed pumping capacities are fully utilized ad pumping costs as a result are low. Regardless of who owns WEMs, the NIS will be distributed in proportion to land owned and the largest WIS will ensure more egalitarian distribution as long as land and labour endowments of families are not directly related.

A third and intermediate situation is where WEM owners are willing to sell water but only at prices considerably higher than pumping costs. Such a situation would arise when water demand is inelastic so that water rent—or, in our terminology, WSS—exceeds the saving in fixed cost which is possible of more water is sold at lower prices. In this situation, wage earners lose since WIS is smaller than in the second situation; buyers of water suffer reduced NIS; and WEM owners are the gainers in WSS. The overall distributional impact would depend upon who the WEM owners are; if land-poor own most of the WEMs, they would claim the bulk of the WSS at the expense of large land owners. But if the distribution of WEM ownership is directly related to the distribution of land then the overall rural income distribution that results will be more unequal than the distribution of land. Notably, concentration of WEM ownership among the land-poor may improve the relative and absolute distribution of gains in their favour under oligopolistic water market structure; however, it may still leave them worse-off than they would be under competitive water markets even if only larger land owners owned WEMs under such a structure.

Communal monopoly is often projected as a superior institutional structure for equitable development and exploitation of a common pool resource such as groundwater. However, in the absence of substantial scalar economies in water production, a communal monopoly by itself may not produce superior productivity or equity results. If, for example, a communal organization establishes its monopoly on all of the community groundwater resource, takes over all private tubewells and sells water to anyone at a price equal to marginal pumping cost to establish Pareto second-best conditions, the productivity and distributional outcomes will be no different than in competitive water market equilibrium. In fact, even if the monopoly charges a price higher than the marginal pumping cost and distributes the super normal profits so earned equally among all members of the community, land-poor families may not be better-off on relative or absolute terms. Where the groundwater resource is abundant, therefore, competitive markets for groundwater may produce best productivity and equity outcomes.

Water Scarce (WS) Areas

Even in principle, the question of distribution of irrigation surplus is far more serious and complex in water scarce areas. Efforts to control over-exploitation of aquifers produce another complex web of distributional effects. Here, we complete our discussion at the conceptual plane of the issues involved in the distribution of groundwater irrigation surplus, assuming that a mechanism can be designed and used to contain the overall rate of groundwater withdrawal by all WEMs together within the safe annual recharge limit.

With total pumping restricted to SAR, the potential for WEM owners to extract oligopolistic rent from water buyers will be particularly great; as a result, the pattern of ownership of WEMs will have a significant impact on the distribution of benefits. However, in WS areas too, emergence of water markets will produce desirable social effects even when they re imperfect.

Enforced autarky will induce and enable owners to WEMs to ignore the 'scarcity value' of water and to use it in a profligate manner, since the opportunity cost of surplus pumping capacity is low. Opening up of opportunities to skim off a part of the new 'irrigation surplus' on others' land through water selling raises sharply the opportunity cost of pumping capacity. In particular, it makes sense for WEM owners to compare the incremental net return fro using a unit of water on own land with the price at which water can be sold to others. In the absence of a market for water as well as of effective checks on withdrawals, the WEM owner owner will use *oa* amount of water on own land. If a regulatory mechanism succeeds in controlling each WEM's withdrawals to *ob,* even in that case, under autarky, the WEM owner will use up all *ob* of water on his own farm although the productive value of water on the margin is low. Opening up of the water-selling option, however, induces the WEM owner to curtail his own use from *ob* to *oc* and to sell water to others so that the marginal value product of water on his own land and the marignal revenue from water sales are equalized. With the emergence of markets, however imperfect, total irrigation surplus that can be produced with the available SAR increases; and the extent of market imperfection determines the mode of distribution of the surplus.

Effective checks on total withdrawal sustain the oligopolistic structure of the water market, for if and as the number of WEM owners in the village increases, the pumping quota of each pumper will decrease. If, however, ways can be found to induce WEM owners in such a situation to behave as if in a competitive water market, then total irrigation surplus will remain unaffected but its distribution may change in favour of buyers.

5

Minor Irrigation Development

Irrigation, being one of the important inputs of agriculture, becomes an equally important component of the rural infrastructure for development. Hence, in any planning for development sufficient weightage has to be given to irrigation development. It is much so in the case of a country like ours where agriculture sector is the main stay of the national economy, accounting for almost half its national income; provides employment for a major part of the population and sustenance to about 70 per cent of it, and yet by and large it depends on the vagaries of monsoon.

The major function of irrigation is to mitigate the impact of inadequate and irregular rainfall, with wide fluctuations from year to year, which often results in even semi-famine conditions. It mainly tries to supplement rain water. But in some cases, it also provides water in area, where rainfall is insufficient for crop production. Irrigation, therefore, not only enhances the productive capacity of cultivated land but also extends the area under cultivation.

Irrigation was known in the country, even in prehistoric times, although substantial development took place only during nineteenth century. The pace was, however, considerably accelerated during the era of development planning.

A characteristic feature of irrigation in India is that quite a significant part of it, is constructed and operated by the State Government. According to the constitution States are responsible for the development of irrigation in their respective geographical areas. As a consequence there is no uniform pattern of administration of irrigation in the States. The Central Government, however, deals with inter-State rivers and in general assists, examines and finances irrigation projects sent by the States.

In Karnataka, agriculture is the main occupation and the means of livelihood to about 67 per cent of working population and contributes about 60 per cent of the income. Out of the total population of about 20 million in the State, 10 million is the working force and about 6.7 million of this are either cultivators or agricultural labourers.

Like other parts in the country, agricultural operations in Karnataka are subject to the vagaries of rainfall. Drought and scarcity conditions are extremely common. It has been assessed by the Irrigation Commission that 75 per cent of the area in the State, has a co-efficient of variability of rainfall of more than 30 per cent and a conclusion has been drawn that Karanataka State clearly falls within the category of areas exposed to a high degree of drought and famine. According to a recent study 127 taluks comprising 143510 sq. kms. in the State of Karnataka, have been delineated as drought prone areas. It is in this context, efforts were necessary to exploit the surface and ground water potential to the maximum extent, so as to provide assured irrigation facility over, as large an area as possible.

Irrigation in Karnataka, as in many other states in the country, received an impetus with the advent of the concept of planning. The development of irrigation in the State during the pre-independence era was rather slow and not uniform.

After reorganisation of the State, a variety of Major, Medium and Minor Irrigation Projects have been constructed and this has considerably increased the area under irrigation. The total area irrigated from all sources was reported to be 7-10 lakh hectares on the eve of the formation of the State in 1955-56 and net area irrigated formed 7.06 per cent of the net area sown. By 1971-72, however, net irrigated area increased to 13-78 lakh hectares, which amounted to 13-3 per cent of the net sown area. There has been a decline thereafter, which could be due to the drought. The peak level attained in 1971-72 could not be restored even by 1974-75. However, an official statement made recently records that the area under irrigation has risen to 19 lakh hectares, or about 19 per cent of area sown, thus registering a substantial increase over the 1971-72 level.

Different sources of irrigation

In order to obtain an idea of the trends of irrigation by different sources we may take up the period from 1956-57 to 1971-72.

It is evident from the statistics given in the Table that during the specified period the net area irrigated by canals and wells has shown over all increasing trend whereas in case of tanks and other sources it has shown declining trend. This may reflect upon priorities given to the sources of irrigation from time of time in the strategies for development.

The development of irrigation has been through massive investments, as it is estimated that from 1974 to 1978-79, over Rs. 773/- crores will have been spent by the Government on major, Medium Irrigation together. This is not inclusive of the loans given to farmers by co-operative and commercial banks for Minor Irrigation. The figures give an idea of the magnitude of expenditure incurred. It is natural, therefore, to expect that management of irrigation should be efficient.

The gross water potential of the State is assumed at 110448 mm^3. Out of the gross water potential of 11044S mm^3 on a rough assessment it is expected that the State can utilise about 50976 mm^3. The total utilization of water to date is placed at about 16992 mm^3. Therefore, there is great scope for further utilisation.

As far as irrigation potential is concerned, Karnataka has yet to achieve the national level of 26% irrigated area in nearly 100 out of 175 taluks. There is still an unutilised potential of about 12-7 lakh hectares under Minor Irrigation and 14-5 lakh hectares unexploited potential under Major and Medium Irrigation. If this irrigation potential is fully developed, the State can make a major breakthrough in Rural Development and removal of poverty.

What are minor irrigation works

Minor Irrigation programmes include a large variety of schemes which differ widely in their technical features and scope. Some require specialised knowledge while others may just depend on local skill and talent for satisfactory completion. The schemes also differ vastly in that some of them are private-owned, others owned by 'community' and some State-owned and managed. Several of these schemes are amenable to public participation in the form of individual enterprise, cooperative venture or public contribution. However, Minor Irrigation works mainly comprise of (a) New works such as Lift Irrigation, (b) Construction of new tanks, including percolation of tanks, (c) Pick-ups or Bhandaras across small streams or rivers, (d) Repairs and improvement of existing tanks, (e) Desilting of tanks and (f) Modernisation of tanks and canal irrigation.

Wells are also Minor Irrigation works, but in Karnataka the exploitation of ground water by source of wells has always been a private enterprise.

From the point of view of administrative convenience and financial implications, all irrigation works that involve Rs. 25 lakhs and less in plain areas and Rs. 30 lakhs in hilly areas are classified as Minor Irrigation works.

The second irrigation Commission of India makes the following observation with regard to the definition of an irrigation project.

"The application of the financial yardstick to schemes constructed long ago was not found practicable, on account of the change in prices. In some cases the cost figures were not even available. In 1959, it was decided that pre-plan schemes irrigation upto 1600 hectares should be regarded as Minor schemes The financial ceiling of Rs. 10 lakhs for individual Minor Irrigation scheme was later raised to Rs. 15 lakhs and has further been raised upward, from April 1, 1970, to Rs. 2.5 million in the plains and to

Rs. 3 million in hilly areas.

The Classification of schemes as Major, Medium and Minor is now a well-established concept in our plans, and may continue both for statistical and administrative purposes. The policy of classifying irrigation on the basis of capital investment does not, however, appear to be rational. With the continuous rise in prices, a work which, at the time of sanction, was classified as Medium, may become Major by the time it is completed. Moreover, the parity between works built at different times is being continuously distributed by the change in prices. We are of the opinion that a physical criterion for classifying Minor Irrigation schemes would be more rational."

Of late, there has been a change in the criterion of deciding a Minor Irrigation work. Financial limit is no longer the criterion. Any irrigation work whose irrigable area does not exceed 2000 hectares is classified as a Minor Irrigation work. It is clear, therefore, that the financial yardstick has been dropped in favor of a 'functional' definition of a Minor Irrigation work.

Over-riding considerations for minor irrigation works

There are definite considerations which make a strong case for implementing Minor Irrigation over Major and Medium Irrigation. Major and Medium Irrigation projects involve extremely heavy capital and operational expenditure. Minor Irrigation, on the other hand costs less to the Government. In many cases part of the expenditure is borne by the farmers themselves. In case of wells the Government comes in the picture as a financing agency along with commercial banks and other financial institutions. But the loans advanced are repayble with interest.

Unlike Major and Medium projects which serve only a concentrated geographical area, the Minor Irrigation works embrace the length and breadth of the State and have a tremendous impact on the development of rural economy of almost each village. It is succinctly opined, "The Primary advantage of minor irrigation is that irrigation facilities provided under such schemes can be decentralized as they are not contingent upon any single major source of surface water. Dry areas which remain almost totally dependent on the vagaries of the monsoon can be served by Minor Irrigation and each farmer can have his own source of irrigation water." Another corollary of this is that Minor Irrigation can serve the small farmer directly, if the accent of planning is on small farmer development.

Further, being small in magnitude Minor Irrigation projects take less time in construction and therefore, bring in immediate results. On the other hand the time required by heavy projects before they become productive is extremely long or in other words gestation period generally becomes quite long.

Judged from a benefit - cost angle in the short run, it is said that the Minor Irrigation works and wells are more beneficent than the Major-Medium Irrigation Projects.

Increasing importance of minor irrigation

The increasing awareness about the importance of and immediate benefits from Minor Irrigation works on the part of the government can be gauged from the investment pattern of Minor Irrigation works.

In addition to the plan programme, there are certain centrally sponsored schemes which are envisaged for intensifying the development work under Minor Irrigation.

The Drought Prone Area Programme (DPAP) was started in the 42 taluks of the State during 1972-73 and so far 929 schemes are completed, costing 453.02 lakhs and benefiting an atchkat of 12,291 hectares of fresh atchkat in addition to stabilisation of 33,996 hectares. There are several other programmes such as integrated development of Western Ghat region and works taken up under the World Bank Assistance.

In the State as per the Minor Irrigation Statistics complied to end of 1.4.1976, these are in all 25,341 Minor Irrigation works. Under the control of the Public Works Department (PWD) benefiting an atchkat of 7,30,730 hectares. There are 15,713 Minor Irrigation works also with an atchkat of less than 4 hectares under the control of the Taluk Development Boards (TDBs). The atchkat benefited by these works is 30,965 hectares.

As regards current development activities, mention may be made of two important programmes, which are undertaken for providing relief in the scareity areas and also in places where the percentage of irrigations low. Under the 15 crores programme it is proposed to execute 1,151 Minor Irrigation works costing Rs. 1,509 lakhs for benefiting an atchat of 33'982 hectares in addition to stabilisation of 50,123 hectares of the existing sanctioned atchkat. Under package programme for Minor Irrigation works with World Bank Assistance, it is proposed to take up 132 fresh works estimated to cost Rs. 798.78 lakhs for benefiting an atchkat of 33,491 hectares in addition to stabilisation of 3,201 hectares. In addition to this improvement works estimated to cost Rs. 1,061 lakhs are aslo proposed to be taken up and completed.

Types of minor irrigation programmes

Administratively speaking, the Minor Irrigation works are executed under the following programmes :

(a) "506 c.o.:" works costing more than Rs. 1.00 lakh are included under this head.

(b) "306 plan" : works costing Rs. 1.00 lakh and below are included under this head.

(c) "Integrated Development of Western Ghats" : This is a special programme which includes certain works specially intended for the region.

(d) "DPAP" : This is also a special programme meant for the development of the area identified as drought prone. There is a special assistance for this programme from the Center to the extent of 50% of the estimated cost of the work.

(e) "Tribal sub-plan" : This is also a special programme which has a substantial population of tribal people.

Since we shall be dealing with the administration of Minor Irrigation mainly under the control of the Public Works Department in the remaining part of this study, it is worthwhile to make a passing reference of the administration of wells, which falls outside PWDs purview.

Administration of wells

As referred to earlier, in Karnataka the exploitation of ground water by source of wells has always been a private enterprise. There has been phenomenal growth in the number of wells in the State in recent years indicating the degree of popularity of well irrigation. As per the Census conducted during 1979, the number of wells constructed from 1st of April 1975 to 31st March 1976 were 4802 wells including 69 bore wells whreas during 1974.75, 6001 wells were sunk in the State.

In the State about 33% of irrigated area is fed by well water. Although the work of sinking of wells is mostly in the private sector, the Government have taken major responsibility for assisting the farmers in utilisation of ground water.

Number of wells constructed in the State

Sl. No.	District	No. of wells constructed during 1975-76	Total No.of wells as on 31-3-1976
1.	Bangalore	208	33,990
2.	Bolgaum	488	46,107
3.	Bellary	232	8,698
4.	Bidar	235	13,558
5.	Bijapur	699	40,139
6.	Chickamagalur	21	1,821
7.	Chitradurga	104	15,361
8.	Dakashina Kannada	610	30,871
9.	Dharwada	247	11,219
10.	Gulbarga	390	14,895
11.	Hassan	55	1,041
12.	Kodagu	2	68
13.	Kolar	430	49,5881
14.	Mandya	127	6,238
15	Mysore	257	8,815
16.	Raichur	267	12,787
17.	Shimoga	48	3,111
18.	Tumkur	225	33,652
19.	Uttar Kannada	156	19,185
	Total	4,802	3,52,144

The State Ground Water Cell of Karnataka was established as a wing of the Department of Mines and Geology in the year 1966. The objectives were to evaluate the ground water resources in the State and suggest a planned programme for development and utilisation of ground water resources to provide assured irrigation facilities to the farmers of the State and thereby raise agricultural production. The cell through its Department is actively associated with institutional financing and rendering technical advice to the public, Primary Land Development Banks (PLDBs) and other Commercial Banks in locating favorable sites for sinking irrigation and drinking water wells. The importance and magnitude of its role can be imagined by taking into account the fact that so far 90,694 such advises have been given. Furthermore, technical opinions regarding the ground water potential and its exploitation by energisation of wells in any given area referred to by the Karnataka Electricity Board (KEB) under REC schemes are being given; so far 156 such reports have been furnished to the Karnataka Electricity Board.

As regards financial assistance, there is provision for liberalised loans for sinking wells. The financial assistance is given to the farmers in the form of loan upto Rs. 5,000 per well.

Karnataka State Co-operative Land Development Bank with its jurisdiction extending to the entire State has a network of 19 branches, one each in the 19 districts of the State. At taluk level there are presently 175 Primary Land Development Banks which disburse the finance made available by the State Land Development Bank to the ultimate borrowers. These Primary Land Development Banks federate into the State level Central Institution. In other words, "the organisational pattern of the long-term credit structure in Karnataka State is of the federal type."

As regards process of taking loans from the Land Development Bank through taluk level branches, the loan applications come to the District Branch of the Bank. Then the loan amount is sanctioned. If the loan amount exceeds the prescribed limits of the District branches, those cases will be referred to the Head Office, Bangalore for sanction. The limit of sanction for the District branches is Rs. 10,000/-. In case of wells, Land Development authorities insist that the area should be cleared by the State Ground Water Cell, i.e., the area should be certified as feasible for digging a well and aslo the State Ground Water Cell should have certified as to the availability of ground water potential to support the scheme.

Land Development Banks advance loans only against the mortgage of land by the farmers. With regard to Commercial Banks, in addition to pledging of land some other security like house, vehicle etc, may also be pledged to get a loan for the construction of a well, and this stipulation is adhered to if the loan amount is not covered.

6

Interdisciplinary View of Irrigation System

Definitions

Irrigation Management has been defined in many ways:

(1) An interdisciplinary system process with built-in learning mechanism to improve system performance by adjusting physical, technological, and institutional inputs to achieve the desired level of output.

(2) The process by which water is manipulated and used in the production of food and fibre.

(3) The process by which agriculture production is increased for each unit quantity of water supplied.

Irrigation management, however, is not water resources dams or reservoirs to store water, nor codes, laws or institutions to allocate water, nor farmers organisation nor soils or cropping systems. It is, however, the way these skills and physical, biological, chemical and social resources are utilised to provide water for improved food and fibre production. Improved irrigation performance depends on the management not only of water but of

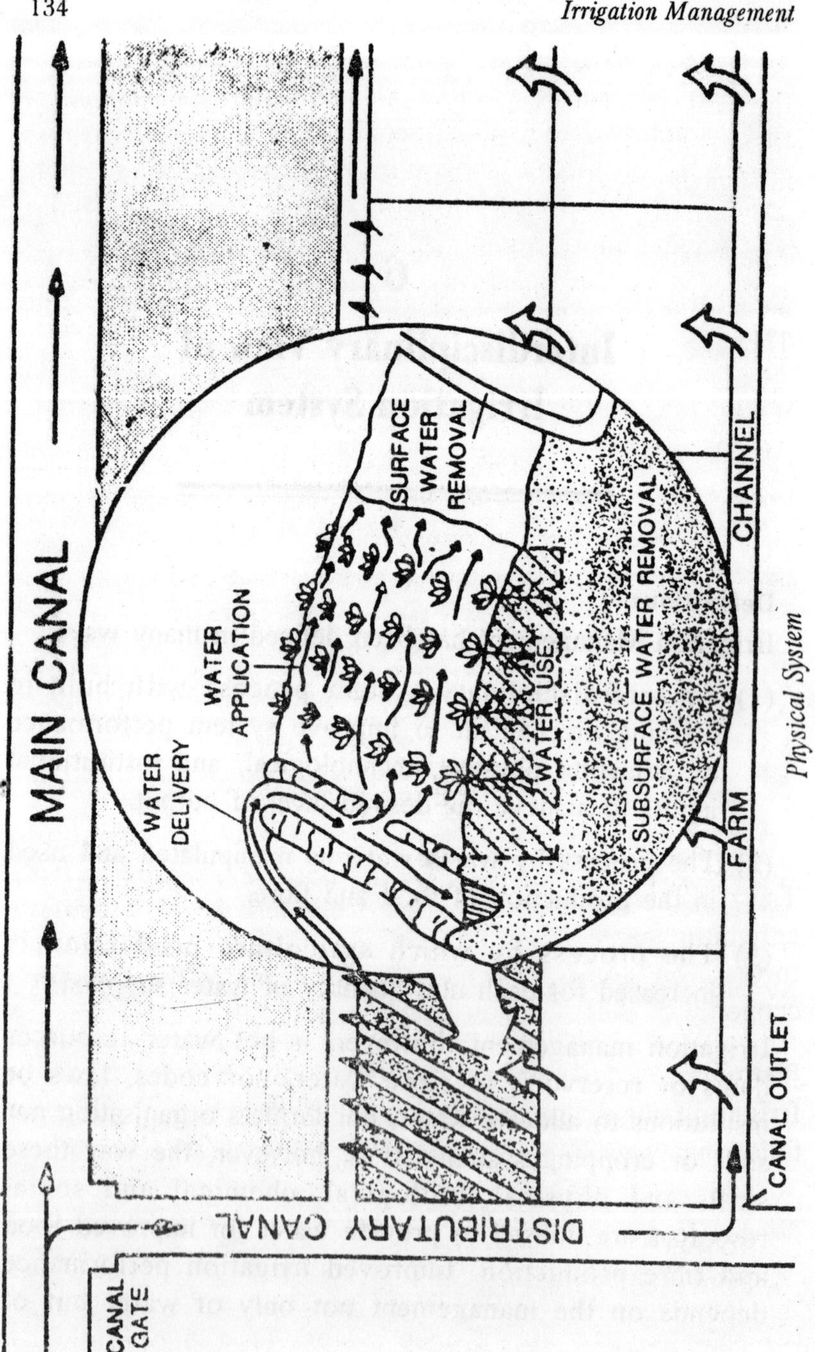

Physical System

irrigation system as a whole, including the management of information and control, of people, both farmers and those who work in irrigation organisations, and of other inputs besides water. Water remains the central though not exclusive focus of attention.

Need of irrigation management

The needs for focusing on irrigation management are:

(1) To conserve water supplies for rapid increase in food and fibre production

(2) To improve the return on investments incurred on existing systems.

(3) To reduce the water-logging and salinity problems which are often mere symptoms of poor management.

(4) To reduce the need for large capital investment in new systems.

(5) To gain knowledge which can provide new criterion for the development and management of other systems.

Therefore the objective of water management in irrigated agriculture is to provide a suitable moisture environment to the crops to obtain optimum crop yields with maximum economy in the use of irrigation water.

Planning

Irrigation management practices can be planned in the following ways:

(i) Advance planning of water delivery with due regard to needs of the crops and engineering aspects of conveyance systems.

(ii) Maintain the conveyance system on top operating

conditions to ensure that allocated water is delivered to the farmers in proper time according to a pre-determined schedule.

(iii) Minimise all irrigation losses, seepage, operational and application.

(iv) Remove from the system excess water that cannot be stored for future use by drainage.

(v) Supply every farmer irrigation-water according to his entitlement equitably and timely to meet his demand as closely as possible within his allocation.

(vi) Adequate measures to improve environmental effects of the project.

(vii) Increased cost effectiveness of investments in the system.

(viii) Adequate monitoring evaluation.

Classification of irrigation system

An irrigation system based on above planning can be grouped into four major components as discussed below:

(1) Physical system,

(2) Soil and cropping system,

(3) Social organisational system, and

(4) Economical system

Physical system

The physical system is primarily responsible for assured supply of irrigation water to meet the crop requirement of its command area for sustained production. It may be subdivided into four classes as figured below and discussed.

Water delivery sub-system

The conveyance of water from the source to the fields through main canal, distributaries, minors and field channels is called a water delivery sub-system. It should function optimally to achieve following expectations

— Flow at constant and regulated rate

— At proper elevation

— Under controlled seepage

— Minimum soil erosion or sedimentation

— Suitable water quality, and

— Proper safety

The water delivery system has been found to be influenced by following factors:

— Flow rate

— Cross section

— Roughness

— Slope

— Seepage rate

— Management.

The water application sub-system should get its due share of water through water delivery sub-system. In case the administrative and executive operational control of the system is in thc hands of government agency or to that authorised by government it must be under defined rules and regulations.

Water application sub-system

The water application sub-system gets its input through

the output of the water delivery sub-system. The former has the following functions :

— Distribution of the designed discharge of water through field channels for field applications

— Satisfy erosion control standards.

— To meet the demand of adequate surface drainage.

— Economically viable and socially acceptable to the management abilities of the farmer.

The following are the factors which influence the water application over the field:

— Filed geometry (length and width)

— Flow discharge rate (water supply rate)

— Slop

— Infiltration rate

— Surface roughness

— Management

The water application sub-system is managed by consumer 'the farmer'. For effective management of the available system he needs to understand three basic requirements of the system:

(1) How to irrigate;

(2) When to irrigate; and

(3) How much to irrigate

Water use sub-system

The water application sub-system supplies its input to water use sub-system. The basic requirement of water use subsystem are:

— To meet the crop water requirements as per soil irrigability class (quantity and quality).

— To maintain favourable salt balance for averting soil-salinity.

— To maintain favourable soil fertility for ensured nutrients level.

To provide favourable soil environment for supporting and nursing of the plants in healthy conditions, avert soil crushing and easy workability.

Water disposal system

The water removal or disposal system contributes for excretion of excess soil moisture from land and thus helps to maintain favourable soil water balance due to which all agricultural operations are otherwise held-up and therefore, cultivation is not possible. The disposal or drainage creates better environment for plants for higher optimal crop production. The excess water to be disposed off from the cultivable lands may accumulate in the fields from rainfall, run off, excessive irrigation, seepage and leakage etc.

The following are the important functions of the water removal sub-system:

— To create favourable air-water balance for better root-aeration.

— To create favourable salt balance within the crop root zone and the entire soil profile.

— To provide conditions for easy tillage operations.

Model of irrigation management

The management of irrigation system is complex. As such

any approach to be adopted for this purpose should, therefore, have all qualities which may cover the system from all angels. The strategy for composition of any model should therefore, have prime object to improve the irrigation system. The development process may, however, comprise of following activities.

(1) Diagnostic analysis

It is a technique by which the benefits and constraints of any system are judged through interdisciplinary approach. The major activities essential for the identification of the problems are:

(a) General overview of the system

(b) Reconnaissance technique

(c) Detailed field investigation technique.

There are elaborated below:

(a) *General overview technique.* This is done by the examining available records and literature pertaining to the system and other investigations carried out in past. As a result of the informations gained, check lists of problems of areas in question and additional requirements to be explored in future should be listed out systematically.

(b) *Reconnaissance technique.* After the general overview of the system the next step should be reconnaissance of farm systems in selected areas, where the investigating team gets opportunity to physically observe the systems to be diagnosed and collect relevant informations from the farmers, related to the problems of physical system and cropping practices being followed. From these interviews and observations, list of problems/deficiencies should be prepared. It is but natural that gaps may appear in the collected informations, which should be documented properly so that additional informations can be recorded during detailed investigations.

(c) *Detailed field investigation technique*. Having acquainted with problems of the area the symptoms are to be investigated by checking several sites of the area for detailed identification. For example, if water-losses in conveyance channels are suspected to be substantial the team should make spot measurements over a short period for several farm systems in supervision of team of engineers. Similarly for other problems such as soil slope, erosion, salinity, soil type etc. are done in supervision of the soil scientist of the team, while all crop studies such as crop type, cropping pattern etc. are done under the supervision of agronomist, the problems of technology transfer and socio-economic conditions by extensionist and sociologist. Additional informations of the areas should not be narrowly defined, otherwise, major problems in the system may be excluded and diagnosis would be incomplete.

(2) Developmental assessment of solution

The various test conducted to study the potential for improving the existing system after listing the solution on a priority basis based on impacts are covered under this phase.

(3) Implementation of programme

This covers the various organizational steps or processes of selection of priorities or impact points for developing a programme of improvement.

The characteristics :

The development model may be classified into three categories.

(i) System perspective

(ii) Interdisciplinary approach

(iii) Farmer's involvement

A system perspective:

The irrigation system is to consist of the following interdependent or inter-related components which may be categorised as under:

(a) physical

(b) cropping

(c) economics, and

(d) social organisational

Miscellaneous classes

A few more classes like legal and others can be added to above classes. However, individually these cannot be evaluated if they are studied cumulatively would only represent the operating irrigation system which is an advancement over previous approach where the problems of irrigation were solely studied or investigated considering it only a physical system.

Interdisciplinary approach

According to the present viewpoint the irrigation system involves components of diverse nature which are certainly inter-related and hence the evaluation through any single or individual discipline with its entire contents of knowledge or experience did not give a complete perspective of the system. This is feasible, therefore, only when the team of interdisciplinary experts works together or simultaneously for resolving the problems and working out appropriate solution for effective management of the system.

The various methods for investigating a system may be following:

(a) The mono disciplinary

(b) The multi-disciplinary

(c) The interdisciplinary.

Farmer's involvement

The farmer who is a consumer and beneficiary of the irrigation water is also a primary manager of the irrigation at the farm level. Hence the interdisciplinary team must bear in mind to have an on-going and vital relationship with the farmers—the user and customer of the system. The general apprehension carried in the mind by experts and planners that farmer's behaviour was responsible for failure of any irrigation system is now not agreed. Today's thinking is that the destructive behaviour of farmers on the system such as cutting of conveyance system, breaking of regulating structures and flow, restrictions etc., might be due to either the defective planning system or lack of appropriate facilities like position of outlets, assured water supply of irrigation, forced the farmer to play with the system which resulted in the failure of the system. Hence the canal offences should be viewed as a compulsion of farmers and as a last resort to erratic and unreliable or un-assured water supply to avoid the crop yield reduction likely to be caused by moisture stress.

The interdisciplinary expert team takes farmers or

group of farmers into confidence and understand his genuine requirement and takes suitable action to satisfy the customer of system. This involvement would certainly reduce the chances of failure of the system.

Field information and data collection

A well designed format for data collection should be developed and carefully pretested in the field for validity and reliability. The information gained should be recorded systematically from available literature, officials, farmers and the selective investigations already conducted during the preliminary survey of the system, farmer problem and identification study. After data collection, procedures are carefully revised and detailed planning of the study is made in the field as detailed below:

(a) Engineering impact factors

These may be divided as follows:

(1) command plan of minor

(2) chak plans

(3) L-Section of field courses channels

(4) land improvement programmes

(5) management practices.

(b) Soil and agronomic factors

(i) Soil factors

These may be listed as follows:

(1) Based on detailed soil survey of the area, the clacification of the soils in different orders, suborders, great groups, families, soils series, types, irrigability classes, salty areas, water table and quality, soil fertility and productivity aspects, physical and

chemical status of soil, soil moisture dynamics and irrigation scheduling based on soil characteristics etc.

(ii) Agronomy factors

(1) Season-wise and crop-wise area in last 3-5 years

(2) Agro-meteorological data of the area

(3) Design-crop pattern as per soil series, type, irrigability class and farmer's preference due to constraints in a system or economic reasons.

(4) Water requirements under different conditions

(5) Weed, disease and other field problems.

(c) Socio-economic factors

The following proforma in local language needs to be developed for recording the informations

(1) Names of farmers

(2) Level of education

(3) Average family size

(4) Average holding size

(5) Availability of agriculture inputs and credits

(6) Marketing facilities

(7) Communication

(8) Social groups

Source of information

The information in a structured questionnaire should be collected by a group of D.A. team in most informal way by taking farmer into confidence through establishing a working relationship. The informations should also be collected from other agencies who maintain regular contacts with the farmers such as village level extension workers, revenue patwaries etc. However, any heresay and surmises should be avoided unless it makes some logical sense. This process would establish a link between officials and farmers and accelerate or stimulate mutual learning, communication and better understanding which is possible through actual self involvement in a recon, naissance exercise.

Allocation of responsibility

The various informations relating to different disciplines would have a collective bearing on the system. The DA team consists of all the experts of various disciplines. The responsibility for collection of various data may be assigned to the concerned experts of this team but the approach for this job should be interdisciplinary and collectively. This is possible only when every member of DA team knows about important activities of other discipline and does not carry out his activities for a particular or own discipline alone in isolation.

Analysis of data and reports

After the diagnostic study has been completed and desired informations or data are collected by the team, these should be collectively analysed and relationship computed for statistical interpretation. The data should be reported in a manner that would define significant problem of area for which solutions are sought. This requires specific ranking priorities of the problems listed out.

Priorities

Ranking priorities are developed for the selection of major problems to find out their solutions depending upon the object. Thus, if the object is to improve agriculture production then each problem or set of problems must be ranked in relationship to the estimated impact they have on crop yields and crop production. However, if welfare priorities are required such as concern for small farmers, each problem must be evaluated in terms of that class and assess how many of each class of farmers are actually having an adverse affect on increased production who need to be helped.

Evaluation problems

Diagnostic analysis technique, determines the key problems which inhibit agricultural production. Having defined the problems, the next step is to search the solutions with the help of area experts, available literature and records etc. The solution of a problem needs to be applied or adopted through demonstration to the farmers. If the problems are specific, applied research should also be carried out to find out the solution of such problems through experiments, tests, or demonstration.

The solutions of problems should be as per conditions of the farmers so as to create the interest of the farmer. This would lead to more trials and lastly adoption of technology by the farmers of the area.

It is necessary to observe that farmer gets institutional and infra-structural support for the diffusion of the new technologies. However, even if feasible or valuable innovation is available but if credit facilities, intensive extension services or other required inputs are not made available to the farmers, they cannot make use of the developed technology.

Communication

Two way communication is necessary to gain feedback

from the farmers about how the technologies are working and is there further need for better improvements for wide diffusion? The decision to release the technology must be made after farmers are well satisfied. However, it is possible through good communication and dialogue in local language between the researcher/extension worker and the farmer for effectively evaluating the technologies being introduced. This assessment is a continuous process from the beginning when various technical solutions to problems are going to recommended.

There should be a continuous sequence of assessment activities, which can be done by the following approach:

(1) The technocrats should discuss the various technical recommendations to be put on trial.

(2) Trial and demonstration by the team of researchers along with extension workers be made at farmer fields to get feedback.

(3) The important tested results should be reviewed for the final introduction of technology.

(4) Follow up action to study the diffusion of recommendations be made carefully from the selected sample of farmers.

(5) A formal assessment should be made about the institutional and resource supply capabilities as needed.

(6) Bench mark studies of some sites surveyed in the problem identification be made to ascertain the impact of the technologies introduced.

Programme implementation

The developed package of technologies which solve

priority problems of area are recommended to farmers for implementation of the programme. This would result in agricultural development with increased food production and improved living conditions.

Institutionalised technology transfer

There should be major focus on the development of institutionalised transfer mechanism that delivers knowledge, resources, and technical assistance to farmers. This helps to fill up the anticipated gap between known technology and implementation of technology.

A programme for technology implementation should be well planned and executed where the necessary organisation is already developed, for which the trained personnel, resources made available to the organisation responsible for implementing the package of technologies with the farmers exists. The evolvement of well developed organisation to execute the planned programme is of paramount importance for the success of the programme.

Diagnostic analysis of irrigation system evaluation techniques

A. Sociology

A. Field studies

Structured interview | Non-structured interview

1. Non-structured interview

An excellent method of gathering certain type of data is the more informal. Non-structured interview helps to ascribe a particular social setting. These interviews are qualitative in nature, meaning that they are not concerned with the members, but with actual processes in a village or Irrigation system.

What is studied

(i) *Life histories:* It involves social experience and studies of the individual as a community member. This involves the farmer about his role as head of the farming family.

(ii) *Cultural patterns:* One could study the values and attitudes of a culture. for example, how does the village culture reels with the issues of tenancy, what values are placed on land ownership?

(iii) Community Organisation : It involves study of rights and dutites associated with different roles in order to understand the structure of relations between people

(iv) Behaviour Patterns : For example, it is common and proper for a large and holder to take more water than small farmers. Does this pattern of behaviour occur at all times? When does it occur?

How are these issues studied

(i) Before entering into the field develop guideline for possible issues and problems to ask about.

(ii) The farmers, officials or organisations to be studied should be clearly defined and carefull selected as representing the general population.

(iii) A preliminary planning is always necessary including defining the objectives and time tables for the study.

(iv) Seek the variety of contact in the community being studied.

(v) The ideas expressed by the interviewer need to be relevant to the person to be interviewed.

Advantages of qualitative field studies

1. It puts the participants in close touch with the system under study.
2. It provides detailed information about the village or irrigation system under study.
3. It reveals the pattern of farmers behaviour.
4. The investigator studies irrigation behaviour in its social setting.

Disadvantages

1. It is likely to affect the farmers behaviour and responses.
2. The investigator takes notes from his own perspective which may unintentionally alter the later interpretation.
3. The overall design can be unsystematic.
4. The unintended selection of community elites as key informants may result in a elitebids in the data.

II. Structured interview: Based on survey questionnaire

Development of the survey questionnaire

The structured questionnaire is the most frequently used instrument for collecting socio-economic data. Understanding its design and its purpose to help the project personnel make the best use of it. It should be so developed so as to get all relevant information in a single interview.

(i) A set of questions to be answered by the respondent.

(ii) Instructions for the interviewer, how to ask questions and record the answers.

Designing the questionnaire

The social scientist is most responsible person for preparing the questionnaire. However :

(1) All team members should be involved in detaining the topics to be covered.

(2) People familiar with the computers should be consulted if the data are to be mechanically tabulated and analysed.

A professional creed

A creed expressing the general attitudes the interviewer should have:

(1) Accept hospitability but never demand it.

(2) Be courteous and respect the people being interviewed.

(3) Personal conduct should have respect for, religious, social and family tradition of the person interviewed.

Content of the questionnaire

The questionnaire should have the following information at the beginning of interview:

1. Name of interviewer,
2. Name of respondent
3. Location of interview
4. Date of interview
5. Time of interview

Introduction

1. How to interview and how the information is collected

2. How the respondent was chosen
3. Names will not be associated with answers
4. Who is sponsoring the survey.

Questions: Nature and sequence

1. Questions should be short and to the point.
2. Questions should use local units of weights and measures.
3. Questions should be understandable and reasonable
4. Closed ended questions should be asked, in simple way which makes the task of respondent easier. For example how many acres of land have you owned? (1) None (2) 1-2 (3) 3-5 (4) 6 or more

1. The open ended questions are usually specific.
2. The sequence of questions should be such that sensitive questions are spread over.
3. The order of questions can add to the quality of the answers as well as increased respondent cooperation.

Conclusion

The following rules help to ensure that the data are both valid and reliable:

1. Cross check the data collected with other key informants.
2. Use different strategies of collecting and analysing data.
3. Use different times, persons, and social situations to provide several accounts of the same event.
4. Use community members to verify the meanings of important concepts.

5. Use personal observation whenever possible to verify, interview data.
6. Use interdisciplinary research team to evaluate interpretations.

Report writing

1. Preliminary section
2. Main body
3. References section
4. Title page
5. Acknowledgement page
6. Preface or forwarding page
7. Table of contents
8. List of tables and figures.

   or we can say all items as INDEX.

1. Diagnostic Analysis for

   Name of the office who is publishing the report
2. Assistance received in collecting the data should be acknowledged
3. It is known as forwarding letter
4. Main heading and subheadings
5. INDEX.

Main body

1. Introduction

Importance of particular survey of existing situation of ideas.

Aims and importance and how it will be benefited to the farmers.

2. Definition of certain problems and technical words.
3. Review of literature
4. Methodology-By which method information is collected for preparation of questions
5. Finding and conclusions.

Reference section

1. *Foot note* : Author's name, Title of report, Place of publication, the year of publication etc.
2. Bibliography : Only to arrange the things
3. Appendix - Index

   1 1/2" should be left towards left hand side.

   1/2" both upper and lower and right hand side should be left.

Collection, information and surveying techniques

Visit, Development, Knowledge, Aptitude communication

1. Accounting method
2. Survey method

   (a) Fill up the schedule, Personal approach

   (b) Scrutiny

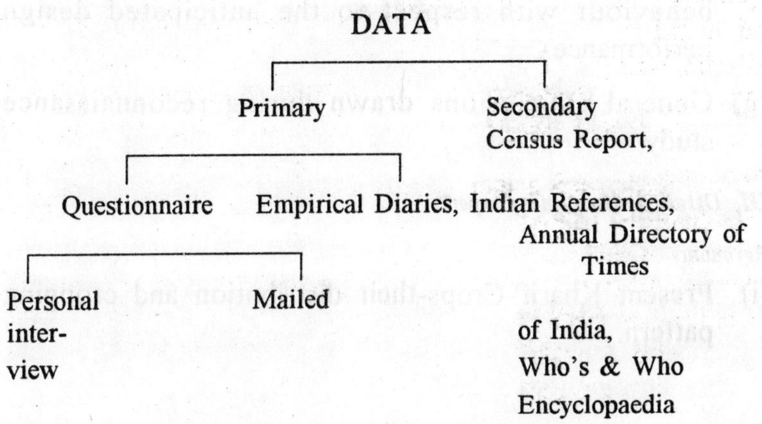

Interview technique

It is a planned conversation is created situation to collect the information to achieve the goal of on-going programme.

General guidelines of report writing of diagnostic-analysis (D.A.) exercise

1. Introduction & objectives

II. Study area

(a) General Description of the area with emphasis on channel under detailed study.

(b) Climatic Conditions: An Annual rainfall, Maximum & minimum temperature; humidity, evaporation, Average sunshine, air etc.

(c) Soils and Ground water—General classificat—ion, soil cover, water table

(d) Crops, cropping pattern and yield

(e) Socio-economic conditions: General condition of farmers, distribution of land holdings, General credit and marketing facilities available.

(f) Irrigation : General status of the system (physical), system of water supply (Rotational/Traditional), behaviour with respect to the anticipated design performance.

(g) General conclusions drawn during reconnaissance study.

III. Detailed disciplinary reports

Agronomy

(i) Present Kharif Crops-their distribution and cropping pattern.

(ii) Deviation of present Kharif cropping pattern from the average of last 3-5 years.

(iii) Expected cropping intensity during the current year considering Rabi and normal.

(iv) General crop conditions, causes of poor crops and estimation of crop yields.

Soils & water

(i) Soils of the area, their classification, physical, hydrophysical and chemical properties, moisture dynamics, irrigation scheduling and irrigation layout plan etc.

(ii) Suitability of crops for the existing soil types as per soil survey

(iii) Water logging, salinity and alkalinity problems in soils and water and their diagnosis and management etc.

(iv) General fertility status and productivity, method of fertilizer application and source of doses used.

(v) Drainage and leaching requirements as per soil problems.

(vi) Water source quality and its impact on soils and crops.

B. Engineering & O.F.D

(i) Main system description—distributary and minors.

(ii) Present status of control and check structures on main system

(iii) Design discharge and deviation therefrom.

(iv) Losses and Conveyance efficiency based on actual observations.

(v) Present status of outlets—type, whether authorized/unauthorized, size, location, installation.

(vi) General condition of water course and field channels, deviation if any from the design standards.

(vii) General field slopes available-whether land leveling or other treatment required.

(viii) Status of farm structures—Drop, check control, division boxes etc.

(ix) Method of application—Wild flooding/border/level basin/furrow or any other whichever applicable.

(x) Method of distribution—Traditional/rotational/demand/semi demand or any other whichever applicable.

(xi) Whether any O.F.D. works done in the command. If yes, their performance, problems/constraints, if any.

(xii) Water disposal sub-system:

Present status of drains, their deviation from the design parameters, their performance, problems/constraints.

(xiii) Anticipated and actual irrigation achieved, short fall if any and reasons thereof.

C. Socio-economic Studies

(i) Socio-economic conditions prevailing at the study site-Average farm size, % distributions with respect to size of land holding, income, average family size, living conditions, infrastructural facilities.

(ii) Status of institutional, agricultural services prevailing:

— Agriculture information received by farmers

— Use and ownership of agriculture implements

— Use of recommended crop practices

— Help received from VLW/contact farmers, if any.

(iii) Status of Institutional services for irrigation:

— Status of Irrigation/farmer's associations

— Status of attitude of operational staff

— Status of maintenance services

— Farmer's satisfaction with respect to the above.

(iv) Status of water distribution with respect to :

— Adequacy

— Reliability/Dependability

— Equity

(v) Present status of resolution of :

— Disputes

— Conflicts.

(vi) Officials opinion regarding operation and maintenance of the system in view of the above observations.

(vii)Cost of various inputs, their source and management :

— Seeds

— Fertilizers

— Pesticides

— Insecticides

— Agricultural implements

— Irrigation water.

(viii) Status of credit and marketing facilities:

— Co-operatives

— Banks

— Local financiers

— Transportation

— Market

Food Corporation of India	Ag. Marketing Boards	Commission Agents	Open market

(ix) Farm Budgeting :

— Gross expenditure

— Gross income

– Net income

D. Conclusions

— General interdisciplinary findings about the system indicating the positive and negative points of the system.

— Problems/Constraints in view of the interdisciplinary approach.

7

Land Reforms in India and their Significants in Equitable Distribution of Water

The imbalance between the distribution of control over land the number dependent on it breeds social and economic inequalities. It also leads to unequal access to developmental activities, institutional facilities, decision making process etc. Furthermore, the existing imbalance offers great social injustice in respect of land tenure system including the sharecropping between tenants and owners. These situations call for an appropriate land reform measure. A more radical the land reform, the greater will be the equity of effect. Several studies made by FAO and other agencies for Asian countries showed the effects of differences between distribution of land holding size and yields. The effects revealed that reduction in either holding size or in equity in land holding distribution is an important factor for increasing productivity and employment. Land reform measure should, therefore, be focused on three basic objectives, viz. (i) higher productivity, (ii) higher employment opportunities, and (iii) equity justice, Thus, an appropriate Agrarian Reform policy is an important strategy for rural development as well as a move for promoting Egalitarian goals for peasants. For balanced and sustained development of agricultural and rural economy, phase-wise transition towards redistribution of land, including other complementary programs in terms of credit, extension services, higher cropping intensity, spread of High Yielding Varieties of crops and overall improvements in technology, it is evident that without effective measures to implement land reform measures and policy for equitable land distribution, it will be extremely difficult to improve the socio-economic conditions of the poor.

Land Reform is not only a political slogan but an activator of technological diffusion and growth in the agrarian sector, obviously, an inquisitive mind is apt to ask why to emphasize land reforms? Japanese had it but, what the necessity of this type of institutional reform? Before we proceed to deliberate on the reforms made in India, it is better to discuss the necessity of land reforms in our country.

Compulsion for land reform, which has happened in countries with widely different historical and cultural backgrounds such as Japan, Iraq. Egypt, Cuba, Bolivia etc., basically emanates from two sources. Firstly, serious supply-demand imbalance in land. Secondly, from the fact that in the initial stage of growth land represent, by and large, the sole source of production in the economy. Ownership and terms governing tenancy vitally affects the life of millions, hence any government gets concerned about the subject.

However, the professional economists neglected the subject of land reforms because it concerns with the institutional framework of the society which they usually preserved as given. Now in the recent past attempts were made to examine the economic base of land reforms. In this attempt the work of N. Georgesca Roger, entitled. 'Economic Theory and Agrarian Economics' which has been examined and extended by Dandekar, is quoted:

> "...defines over-population to be the situation in which, under conditions of full employment, the marginal productivity of labour falls short of minimum subsistence of the worker. In the more extreme situation, the marginal productivity of labour may even reach zero before the conditions of full employment are realised.... Most underdeveloped countries do show such signs of overpopulation in a more or less extreme form. The process of economic growth consists in accumulating more and more land capital resources so that, marginal productivity may progressively rise—first above the zero and then above the minimum substance. It is then that the condition of overpopulation ceases to exist."

From the standpoint of land reforms, Dandekar continues to elaborate Roger's approach (quote):

> "It should be recognized further that in the countries concerned, the agricultural sector is usually much more acutely over-populated than is the economy as a whole. This is because the non-agricultural sector in these economies is usually organised on the capitalist principles and hence does not permit workers in, unless they can contribute to the production more than the wages, they received in return. Consequently,

> the entire residual population is thrown on the agriculture which by its nature and tradition employs or accommodates whatever population is thrown on it without reference to the marginal productivity of labour conceived as a part of the problem of economic growth, the agrarian problem consists in holding on this population until an increasing part of it is withdrawn to the non-agricultural sector and in the meanwhile in employing it usefully so as to maximise the total output of the agricultural sector.... That capitalism cannot offer solution to this problem is easily demonstrated. Because capitalism, understood as an economic system regulated by profit maximisation, cannot offer employment to labour beyond the point of its marginal production, therefore remains unemployed. Apart from its social consequences, this is obviously not even an economic solution for, though it maximises the profit—rent of the capitalist entrepreneur, it fails to maximise the total output of the agricultural sector."

Viewed from this angle a non-capitalist type farming, or conversely speaking individual peasant holding, offers opportunities though at a wage rate, less than the marginal productivity to a very large number of farmers—family workers etc. Thus, land reforms promote maximisation of total output that has an economic rationale in a typical economy that LDC's represent.

Significance of land reforms in India

1. Promotion of Incentives to invest and earn more by which process of capital formation can be accelerated.

2. Social Justice : Tiller of the soil be given an economic status, Concentration of economic power that a large ownership of this productivity resource represents is not conducive for the growth of democracy. Therefore, land to the tiller is a motto to obtain social justice.

Viewed from this standpoint, the concept represents two institutional reforms : one is related to agrarian relations and the other to the size of the unit of cultivation. The latter component has quite often been objected upon the arguments, will not this redistribution of ownership reduce productivity by dividing up large estates? this is really a very important question and hence, it is proposed to examine it later in the context of land ceilings as a measure in land reforms in India.

In different countries land reform measures have been introduced to meet different requirements of individual countries. According to U.N. Survey, the programme of utilization of land incorporates several reformative policies. In Canada and Australia, it has been done with the basic objectives of conferring ownership right on tenants. In Spain it has embraced a much wider field with a view to improve both the economic conditions of farmers and output of the farm. On the other hand in Italy, land reforms include measures for the improvement of the social condition both for moral and civic reasons as also for economic considerations. The objectives set forth in the Agricultural land law of Japanreads out, "in recognition of the fact that agricultural land shall be most appropriately owned be none other than the cultivators thereof."

If all these land reforms discussed above are to be

catalogued into suitable categories, we would notice that either one or more of the following aims intended to be achieved:

(i) To provide an opportunity for the development of farmer's personality. In agrarian economies land has economic and prestige value. Without possessing land of their own, even a small piece of land, people have no status and suffer from a sense of injury.

(ii) To Provide an opportunity for the development.

(iii) To climinate the exploitation of one class with another:

(iv) To remove the inhibitions at institutional level towards augmenting agricultural output, both total and per unit of land.

(v) To increase the stabilise the volume of farms employment through a wiser distribution of the ownership of land.

(vi) To bring about a fair and equitable distribution of agricultural income.

Land reform in India represents a grand design to revolutionize the relations between the government, the owner, and the cultivator of the land, change the size of farms in an economically meaningful way and thereby revolutionize investment technology and productivity in conjunction of course, with other major policies. However, we would be benefited by a reference to the then Prime Minister's Inaugural Address to the Chief Ministers' Conference on Land Reforms on 26th-27th September, 1970 (quote).

> "…. During this period, we have had unmistakable evidence, and clear warning of the awakening and impatient in the country size... the people in this country have entrusted their destiny in our hands. We owe it to them to think deeply and honestly over recent developments. Why is social discontent in the country side on the increase? Why is it erupting into violence with greater frequency? The answer to these questions is obvious, but is bears repetition. The land reform measures implemented have failed to match the legitimate expectations which were first fostered amongst millions of cultivators during the national movement, we have yet to create an equitable tenurial system for the village poor….in short, we have yet to create institutional condition which would enable small-farmers, tenants and landless labourers to share in the agricultural New Deal."

We have quoted this paragraph in order to highlight the reasons, of course as visualised by the party in power i.e. the government, why land reform has become a problem of national importance and national urgency. Concurrently, it admits failures in a very striking manner. As a student of economics, however these reform measures have to be understood with academic coolness and political overtures.

The First Five Year plan set out a broad outline of the policy to be followed by the State governments; the policy was elaborated in the Second plan:

(i) to remove impediments in the way of agricultural production as came from the character of the agrarian structure and to create conditions for evolving as speedily as possible an agrarian economy with high levels of efficiency and productivity.

(ii) To establish an equalitarian society and eliminate social inequalities.

The Third Plan envisaged the implementation of the policy evolved during the earlier plans, which period is characterized by legislative measures on the various aspects. At the time of the Fourth Plan it was realised that the entire policy need to be oriented having regard to the technological developments in agriculture and the social requirements of the time, is a reflection of the pressing need to be effective and realistic. To review the progress of land reform measures, to identify weakness and gaps in the land laws and their implementation and to review the land policy in the context of technological developments in agriculture-a Central land Reform Committee was set up in 1970 with Union Minister of Agriculture as its Chairman. As a result of this, reoriented policy and directives from the Committee, and ceiling land redistribution was accelerated.

The draft Fifth Plan observed: 'The objectives of land policy have been to remove such non-motivational and other, impediments to .... increase.... agricultural production as arise from the agrarian structure inherited from the past and also to eliminate all elements of exploitation and social injustice within agrarian system so as to ensure equality of tenurial status and opportunity to all sections of the rural population'.

The various reformative measures when pieced together reveal that reforms of the following types were introduced:

(i) *Abolition of Intermediaries* : This toped the list in priority. This system of Zamindari/Jagirdari/ Talukdaries and in all covering 40% of the area of the country had to be done away fortwith. Owners of large holdings resorted to dilatory tactics through courts. However, this could be through. Legislative provisions, however, had yet to be enacted (1970) for abolition of some of the remanants of intermediary tenures in some states such as temporarily settled estates in Assam, intermediary interests in Tata nagar, Zamindari in Bihar, Sripandar Vogai lands in Kerala etc. The main problem faced by the government is to bring abut uniformity in the rights of tenants and rationalisation of assessment payable by them, and the build-up of effective land revenue and them, and the build-up of effective land revenue and land reform agencies in the former zamindari areas.

(ii) *Tenancy Reforms* : Three types of cultivating tenants existed. These were Occupancy tenants, non-occupancy tenants, i.e. tenants-at will and sub-tenants. These continued even after the abolition of intermediary tenures. Various measures have been taken to give security to the tenants; fixation of fair rent and eventually to bring them in direct contact with with state. Further, the right of resumption by the land owner from his tenants has been further curtailed.

(iii) *Ceiling on Land Holdings* : The entire range of problem relating to ceiling has been covered at the later stage, in a way continuation of this.

(iv) *Consolidation of Holdings* : Indian farming was known for its absentee landlordism and for smallness and fragmentation all over the world and evidence of this is available in the Report of the Royal commission on Agriculture in India. Consolidation, unlike other land reform measures does not relate to tiller-owner dimension of the problem. However, it proposes to do away with the scatteredness of the operational holdings. With varying degrees of success each State in India has gone ahead with this programme.

Impediments in Implementation : Implementation of land reforms is beset with numerous difficulties: These are political, social, economic and legal. Here, only the more significant ones have been listed.

1- *Defective Laws* : Legislative measures adopted from time to time contained many exemptions and allowed flexibility to states and thus the entire aim of putting the agricultural land within the orbit of land reforms has been defeated. Some argue that his element was an off-shoot of the lack of political will to radically alter the tiller-ownership relationship.

2. *Inadequate Application* : Time lag between the intention to alter the existing land-owner-tiller relationship, enactment into laws and its application has often helped the owner to make fake transfers taking out the land from the legal ambit. This trend got further incentive from the inadequate administrative support. Regarding implementation and the required coordination we may quote extensively with advantage, from the 5th Report on the Progress on Land Reforms of the United Nations.

".... While on the one hand the nature of the land reform programme would indicate the nature of the administrative organization required for its implementation, on the other hand administrative factors such as the nature of the existing administrative structure, the availability of personnel, the existence or otherwise of land records would dictate the scope, timing and phasing the programme. For instance, an integrated programme would require large number of trained personnel in a number of different fields. In the absence of such personnel, it would be useless (and even harmful) to introduce with a programme, since it would be deemed to failure. Similarly, the granting of individual ownership would require and adequate and accurate system of land records... the production of which through cadastral survey and land registration would take a long time".

After Independence the rural sector remained under a bureaucracy meant originally to rule and not serve; and records were anything but adequate.

These two deficiencies in conjunction with class-caste ridden social structure aggravated the situation of non-implementation. Under the constitution, the main responsibility for enactment of land legislation and its implementation is that of the state governments wherein local pressures are bound to offer difficulties. Uniformity received strength after the meeting of the Central Land Reforms Committee (CLRC) held on 3rd August, 1971, under the chairmanship of the then Minister of Agriculture.... Hence, it is proposed to give their main recommendations later in this Chapter.

Thimmaiah and Aziz outlined the causes of failure in implementation of Land Reforms in Karnataka which are :

Caste System and its interaction with State power and bureaucracy. Caste System contributes to the failure of Land Reforms in two ways (i) : It weakens class consciousness among the potential beneficiaries and inhibits growth of a supporting peasant movement, and (ii) Caste alliances of the rich landowners with bureaucracy, and those in State power work against the effective implementation. Further, the role of Caste System in contributing to the failure of Land Reforms has been highlighted. The Inam Abolition is explained in Caste Terms. The owners of Inam lands were mainly Brahmins, absentee owners, and the tenants of these lands belonged to the dominant majority castes who constituted the vote banks after independence. It suited the political interests of those in power i.e. the no-Brahmins landed interest viz. Vokkaligas and Lingayats, to bring in legislation for abolition of Inams. But the bureaucracy was controlled by Brahmins, and this contributed to the delay in implementation. Though, legislation for abolition of Inam lands was passed in 1947, it was made into an Act only in 1954.

If the delays in the implementation of Inam legislation are attributed to Brahmins bureaucracy, which had a vested interest in Inams, the delays in initiating a radical Ceiling Legislation upto the early 1970's are attributed to the dominance of the land owning Lingayat group in State power. The growing power of the dominant non-Brahmin caste groups is illustrated by the facts on the changing composition of the Karnataka Legislation Assembly by caste. While the share of Brahmins in the Legislative Assembly declined since 1952, the Lingayats and Vokkaligas held dominant position in the Assembly with more than 60% of the Legislature. There was also a shift from non-agricultural to agricultural groups. Thus, the dominance of non-Brahmins land-owners in State power contributed to delays in radical ceiling legislation while resulting in a stiff tenancy legislation, the contrast is traced to the interplay of caste politics. While ceiling legislation affects the land owning Vokkaligas and Lingayats, the tenancy legislation affects the Brahmin absentee landlords and the Vokkaliga-Lingayat coalition in State power had less to lose by a stiff tenancy legislation. The more radical ceiling legislation of the early 1970's is attributed to the installation of non-dominant minority backward group in power, reduced the strength of dominant landlord caste groups in legislature and the support of a Brahmin bureaucracy which by this time lost its landed interests.

As regards the implementation of the 1974 Act, the post 1977 period implementation suffered a setback as compared to the pre-1977 period. This is explained mainly in terms of a number of factors; the renewed power of the dominant majority caste group, the growing share of the dominant majority caste groups in bureaucracy and the conversion of land Tribunals into an institution for distribution of spoils among politicans.

Ceilings on land holdings : The controversy

Emphasis on ceiling on land ownership underwent metamorphic change between the 1950's and the 1970's some lessons appear to have been learnt from the past. The policy of ceiling would be able to make contributions of the following types.

1. Meeting the widespread desire to posses land
2. Reducing glaring inequalities in ownership and use of land.
3. Reducing inequalities in agricultural incomes.
4. Enlarging the sphere of self-employment.

While obtaining the above mentioned results, it has to be ensured that the ceiling does not lead to the establishment of uneconomic holding: in other words the size must be efficient to adopt and work under the given technology with a degree of efficiency which should not suffer with size disabilities.

In this context, that is the yardsticks... for determining the minimum farm size, three norms are available, a bullock unit, a labour unit, and an income unit; ceiling may be a multiple of such holdings. Usually, three types of the family holdings (determined by any one of the three norms listed above) was recommended. Let us see what these three norms imply:

(a) *Bullock Unit* : The family farm should be of a size as to employ fully the indivisible inputs especially a pair of bullocks.

(b) *Labour Unit* : The family farms should be of a size that it would employ full the labour of the farm family (manpower of an average sized family) without the need to work outside the farm.

(c) *Income Unit* : It should provide a minimum income for a civilized life, for example Rs. 3500 per annum per family.

On a deeper probe, it can be established that farm size should be determined by any of the above noted yardstic which will vary from region to region and in the same region according to the level of technological development.

Whatever may be the yardstick, a nation characterised with deficits in most of the products, the ceiling on land holdings has to be such as not to introduce size disabilities. Thus, the issue of maximum permissible ownership is linked with the minimum size since ceilings viewed as an aid to production. The Congress Agrarian Reforms Committee defined an economic holding as one, under the given agro-economic conditions, that would provide a reasonable standard of living in addition to full employment for a family of normal size and atleast a pari of bullocks. The planning Commission replaced the norm of economic holding with that of a family holding equivalent to a plough unit/or to a labour unit working with such assistance as is customary in agricultural operations. The exact area of a family holding was to be fixed in each region separately according to various factors such as type of soil, nature of crop etc. In may, 1955 the Planning Commission set up a Panel on Land Reforms which recommended : "The limit should be three family holdings for an average family in which the number of members does not exceed five. Due to additional family holding should be allowed for each additional member subject to a maximum of six family holdings.

Since, no central direction was given prior to 'Central Land Reforms Commitee', each state let free to adopt its norms fully utilising the defining factor of the local condition with the result that:

(a) In some States family was taken as the unit and in some the individual.

(b) Relative valuation of different types of land (irrigated and unirrigated land) was tackled differently.

(c) Various kinds of exemptions were provided in the name of efficiently managed farms/plantations etc.

(d) Under different laws transfers were disregarded differently i.e., in Andhra Pradesh, Uttar Pradesh,

Himachal Pradesh and West-Bengal, all transfers, after the prescribed date were null and void or disregarded, but in Bihar, Delhi, Manipur and Tripura transfers were made through registered documents for valuable considerations.

State	Level of ceilings (Acres)	(hectare)
1. Andhra Pradesh	27 to 324 acres	10.93 to 131.2
2. Gujarat	19 to 132 acres	7.69 to 53.44
3. Mysore (now Karnataka)	27 to 216 acres	11.74 to 87.45
4. Rajasthan	22 to 336 acres	8.01 to 136

No doubt that effective action under land ceilings could not make much headway which can be ascertained from the following details.

Sl. No.	*State*	*Surplus area (in acres)*	*hectare*	*Area distributed (in acres)*	*hectare*
1.	Andhra Pradesh	73692	(29834.82)	Nil	
2.	Assam	67934	(27503.6)	466	(188.67)
3.	Gujarat	41030	(16611.34)	14000	(5668.02)
4.	Haryana	182250	(73785.43)	54981	(22259.2)
5.	Kashmir				
6.	Madhya Pradesh	75581	(30599)	12500	(5060.7)
7.	Maharashtra	262741	(106372)	116260	(47068.8)
8.	Punjab	191527	(77541)	60333	(2.4426)
9.	Tamil Nadu	25229	(10214)	15542	(6292)
10.	Uttar Pradesh	140554	(97390)	123588	(50035.6)
11.	West Bengal	794410	(321623)	182338	(73821
12.	Himachal Pradesh	6525	(2642)	292	(118.22)
13.	Tripura	148	(60)	Nil	
14.	Delhi	405	(164)	Nil	

Recommendations of the central land reforms committee

The Central Land Reforms Committee gave a new dimensions to the programme of land reforms by recommending certain uniformity in ceiling laws plugging many of the known loopholes : it recommended a lower ceiling unit. For fixing the norms were clearly laid down, exemptions were taken-out and more significant is their recommendation to fix ceiling in various states within certain limits. In the following paragraphs we give their recommendations regarding the ceiling on land holdings as finalised in their meeting held on 3rd August, 1971:

(i) Ceiling should be applicable for the family, as a whole, the term 'family' being defined so as to include husband, wife and minor children.

(ii) Where the number of members in the family exceeds five, additional land may be allowed for each member in excess of five in such a manner that the total area to the family does not exceed twice the ceiling limit for a family.

(iii) The ceiling for a family of five member may be fixed within the range of 10 to 18 acres of perennially irrigated land or irrigated land capable of growing two crops. Since, agro-climatic conditions vary from area to area, the committee only laid down the range within which states were free to fix the ceiling.

(iv) For various categories of land ; conversion-ratios should be fixed taking into account availability of water, productivity, soil classification, crops grown etc. However, an absolute ceiling for a family even in the case of dry lands was put at 54 acres thereby limiting seizable variations.

(v) Exemptions in the existing laws in favour of mechanised farms, well managed farms, etc. were to be withdrawn.

(vi) Exemptions in favour of plantations of tea, coffee, cardamom, rubber etc. should be carefully examined in consultation with the Ministers concerned and State Governments. Thereafter, this and other types of exemption should be discussed with the Chief Ministers in order to formulate the national policy.

Since then, that is after the Central land Reforms Committee's recommendations and consequent changes in the socio-political atmosphere, ceiling laws of most states have been amended and the present position with regard to the ceiling is indicated (as illustrated only a few states have been listed) in the Table to follow :

Sl. No	State	Level of Ceiling (hectares)
1.	Andhra Pradesh	4.05 to 21.85
2.	Bihar	6.07 to 18.21
3.	Haryana	7.25 to 21.85
4.	Himachal Pradesh	4.05 to 21.14 (in certain specified areas upto 28.33 hectares
5.	Punjab	7.00 to 21.80
6.	Uttar Pradesh	7.25 to 18.21

The total land donated to the Sarvodaya leader, who had launched the 'Bhoodan' and later the 'Gramdan', still remains undistributed and thus, 24,40,947 acres of land awaits distribution.

According to latest estimates, 22,42, 424 acres of land has been distributed till now, now out of the

46,88,371 acres received in 'Bhoodan'. A total of 24,40947 acres remain to be distributed.

Bihar, which had led in the land donation movement with a tally of 21, 19,670 acres, is the major defaulter with 14,92,403 acres still to be distributed. However, out of the more than 21 lakh acres donated during the mid-fifties, only about 8,49,267 acres were found to be fit for cultivation. Of this, only 6.27,267 have been distributed and about 2,22,000 acres of culturable and 12,70403 acres of unculturable land is still to be distributed.

The next major defaulter is Rajasthan. In this State, out of the 6,01,219 acres received in Bhoodan, only 1,40,496 acres have been distributed among the landless. According to the report furnished by the State Government, distribution work was in progress in the Chhattisgarh area in Bikaner district. this area, it may be mentioned, has recently come under the command area of the Rajasthan. Canal project. Either, it was all more or less desert. The State Government has also reported that records of 50,000 acres out of the donated land were being verified from the Revenue record.

Although, the donations in Gujarat were modest—only 1,34,457 acres—the distribution too has been very poor—only 27,131 acres.

Out of the 4,10,239 acres received in Madhya Pradesh 1,67,720 acres are still to be distributed out of which about 57,000 acres are said to be unfit for cultivation.

Orissa appears to have registered the best achievement with 5,71,970 acres out of the 6,33,864 acres received in bhoodan. The reconstituted State Bhoodan Yogna. Committee, has decided to distribute the remaining land within three years and to plant trees etc. on land found unsuitable for cultivation.

In Maharashtra, a total of 1,09,999 acres were donated, out of which 83,380 acres have been distributed.

Uttar Pradesh has achieved the best results., better than that of Orissa even of the 4,36,421 acres of land donated, ad much as 4,20,802 areas have been distributed.

Small areas were donated in son. other States too, like Assam, Haryana, Karnataka, Punjab and Tamil Nadu.

Size and productivity : But for the recent years a great debate followed, during the Second Five Year Plan and immediately following years, concerning the size desirability that would be introduced due to lowering of the ceilings. At the back of the argument was the commonly accepted notion that size and productivity are highly correlated dimensions and any reduction in size would, if so affected mean,lowering the productivity of farm efficiency. Fortunately the Farm Management Survey, initiated in India on a national scale in 1954 that this politically sensitive issue could be examined scientifically.

It would not be feasible to attempt a detailed review of each one of the arguments for and against size and productivity relationship. However, a reference to the below noted literature will be rewarding to those who desire to pursue this line of debate further :

1. Dantwala, M.L. Impact of Redistribution and Pooling of Land on Agrarian Structure and Efficiency of Resource Use; Indian Journal of Agricultural Economics (IJAE), Vol. XIV, No. 4, Oct. Dec. 1959 Bombay.

Incentives and disincentives in Agriculture. Indian Journal of Agricultural Economics, (IJAE) April-june, 1957.

2. Krishna, Raj — Firm and Optimum Farms', Economic Weekly, October 31, 1962.

3. Rai, C.H. — Hanumantha 'Farm Size and Economics of Scale', Economic Weekly, 14 December, 1963. pp. 2041-44.

4. Sen, A.K. — Size of Holding and Productivity, Economic Weekly, Annual Number, 1964.

5. Khurso, A.M. — "The Economies of Land Reforms and Farm Size in India', published by Macmillan India (1973).

Without going into the methodological series, these studies by and large reveal an agreement in academics on certain issues:

(i) Farm size can be defined either in terms of a single input, say acreage or in terms of output, farm efficiency either in terms of output per unit of a single input. acreage or as output per unit of cost of all inputs. Both these sets suffer with limitations and a policy makers have to take a decision in the light of yardsticks that are intended to be used.

(ii) As farm size (acreage or hectare) expands, gross output per acre declines. There are certain studies contributing to the idea of constant returns to scale.

In any case, not many have disputed this contention. To quote from the 'Studies in FArm Management in West Bengal, Report for 1954-55 to 1956-57" (Economic Weekly, Annual Number, February, 1961, p. 326).

"Farms of moderately large size rather than the biggest ones are observed to have earned the maximum profit. The clue to the successful performance of these farms lies in the adoption of multiple and diverse cropping pattern which combines in a satisfactory manner growing of different crops. The farms that have earned the maximum profit out of all the farms in the three years have farms size 9 and 12 acres.... usually the farms of moderate size are observed to earn maximum profit per acre while it is mostly the small farms that have incurred very high loss per acres. (Explanation : This loss has been ascribed to total or partial crop damages due to drought and flood).

(iii) As farm size expands (acreage), net profit per acre increases which stands in sharp contrast to that of decline in productivity with increase in size. This is so because small farms have a surplus in overheads/home labour and when a shadow price is placed such home produced inputs per acre cost goes up and their profitability (not the yield per acre) goes down.

Taking a broader view of the situation the impression of constant returns to scale is quite strong. If such is the case, then it dispels the fear that land ceiling can adversely affect agricultural productivity efficiency. Of course, below a minimum, the ceiling may adversely affect the 'Cut off' point which has to be kept in view.

With the introduction of seed-fertilizer-irrigation break-through, usually captioned as the "Green

Revolution" the High Yielding Varieties Programme (HYVP), the possibility to further lower down the ceiling has gained impetus. For, some time, it was feared that the adoption of 'New, Technology' may get a set back due to the recommended lowering of ceiling by the 'Central Land Reforms Committee. The All India Survey was launched to adversely affect the productivity efficiency of farms consequent to the lowering of ceiling on farm holdings. On the contrary, indications are that the New Technology carries potent of radical lowering of ceilings in regions endowed with better Infrastructure i.e., endowed with an assured irrigation/financing facilities. For empirical evidences, we may refer to various studies offering on the subject from time to time including those listed below :

1. Conference Number of the Indian Journal of Agricultural Economics, Vol. No. XXVIII. No. 4, October-December, 1972.
2. National Council for Applied Economic Research : "Impact of Mechanization in Agriculture and Employment" (1973).

Thus, the present day controversy on land ceiling is more on the inter-sectorial income parity i.e. the level of income that is agriculurist has the moral right to claim to earn from his land, keeping in view the level permissible to his urban counterpart (entrepreneur).

Before concluding this, let us be clear that importance of ceiling on farm holdings without care to intersectoral balance issue, and to place it below a critical minimum may create serious economic and social problems. But, to keep theceiling too high, will forefeit the main objectives of the land reforms. Effective carriage of the philosophy of 'land to the tiller' plus adequate infrastructure facilities specially suited to small farmers like adequate provision of irrigation, fertilizers, need and cash finance and linked with the facility of efficient marketing institution are the most urgent needs of the nation. Thus, the programme of land reforms has to be an integrated business and should not be viewed in isolation.

Role of land reforms in equitable distribution of water

The average size of holding in India is as low as 2.3 hectare and that too is handicapped with scattered plots/ fields of the farmer's. Besides, the condition is further, aggravated on fragmentatoin on holdings of the farmers. Further, the imbalance between the distribution of control over land and the number of dependents on it breeds, social and economic inequalities. It also leads to unequal access to irrigation water more particularly to surface water, i.e. canal, and also to developmental activities, institutional facilities, decision making process. Moreover, caste system also contributed to the failure of land reforms in the country and rather helped in unequal distribution of the ownership of land.

The failure in the redistribution of land to the poor through land reforms is well documented in several earlier studies by the erstwhile Research Programmes committee of the Planning Commission, Government of India. This has also been the theme of a number of other subsequent studies which have also revealed the failure in the redistribution of land to the marginal and small farmers through land reforms.

Implementation of land reforms could not successfully be affected with the seriousness it was expected. Therefore, land reforms could not contribute much in equitable distribution of water, which mainly happened due to:

(i) Defective land laws

(ii) Inadequate application,

(iii) Smalless and fragmented holdings,

(iv) Improper Land leveling and shaping, and

(v) On-Farm development.

Legislative measures adopted from time to time allowed flexibility to States and thus, entire efforts of consolidation of holding have been defeated. The purpose to do away with the scatteredness of the operational holdings and put it into one or two compact blocks is also lost due to detective laws of inheritance. Since, after every 10 to 15 years, the problem of scatteredness and fragmented holdings would arise. Such a situation does not contribute much in equitable distribution of water. A farmer with 3 or 4 chaks of his holdings can neither have tubewells or boring in each chak nor he (farmer) can afford to have expensive irrigation i.e. small scale lift technology.

8

Why Farmer Participation

Farmer participation is presently more often a potential than a reality. In agency-managed systems, the users' role is usually restricted to activities "below the outlet,". Even in user-managed systems, the amount and effectiveness of participation can be less than desirable, for example, if decision-making is dominated by rural elites, or if water distribution does not reach all tail-end farmers. It is important to be clear about the goals of irrigation management, to have some criteria by which to judge when more or less participation may be desirable, and also how much and what kinds? Furthermore, the benefits and costs need to be considered, preferably in relation to one another, though there are few analyses which permit such comparison.

Objectives in irrigation management

The goals that may be furthered by farmer participation in irrigation systems management are seldom of equal importance to all concerned. The things sought by farmers and by government agencies may or may not be congruent. Moreover, farmers may disagree among themselves on the weighting of objectives, as may government agencies, according to their responsibilities and interests. some goals, fortunately, may be similarly appreciated by farmers and officials.

The objectives themselves are often interrelated and cannot be ranked according to some fixed priority. From the literature we can identify the following criteria, some instrumental to others. The first and last categories represent the broadest goals, whereas the others are more specific and sometimes intermediate. all can be furthered by farmer participation in various ways.

Greater production or productivity: measured either as total output or as the amount produced per acre or per unit of water, to be achieved through some combination of increases in:

1. Yield,
2. Area cultivated, and/or
3. Cropping intensity, i.e. more crops produced in a year.

Such increases can come from more adequate or more timely water application, from savings of water within or between seasons, or from use of new technologies, such as higher-yielding varieties or fertilizer and agricultural chemicals, made possible or more profitable by better water management.

Improved water distribution, which has two aspects:

1. Greater reliability and predictability in the amount and timing of water deliveries; this can encourage use of new deliveries; this can encourage use of new technology and make possible more efficient use of labor.

2. Greater equity of distribution, particularly between upstream and downstream areas; this can contribute also to production and productivity.

Reductions in conflict

1. Among users, e.g. between upstream and downstream farmers, so that cooperative water use in more possible.
2. With government agencies, so that main system management is less impaired or less subject to political interference.

Greater resource mobilization

1. Contributions of labor and materials:
   a. during construction or rehabilitation,
   b. for system operation, at lower levels and upwards as established by agreement or precedent, and
   c. for system maintenance (routine, preventive and/or emergency), at lower levels and above.
2. Contributions of funds :
   a. toward the capital costs of construction, and
   b. toward the recurrent costs of operation and maintenance.
3. Contributions of information :
   a. during design activities or rehabilitation planning, and
   b. for improved operation and maintenance.
4. Cost and quality control by farmers such as inspection of work by contractors or agencies, scrutiny of

materials delivered for use, and overseeing work at field level.

Sustained system performance

1. Managing soil and water resources so that their productivity maintained; for example, too little or too much water application can lead to salinity, depending on soil and water table conditions.
2. Achieving and continuing intensified production so that a larger population base can be supported.

Production objectives probably command the broadest support among both farmers and government agencies, though governments tend to look at aggregate levels of production. Farmers, on the other hand, are more concerned with who is getting the production increase and with what net benefit. The latter is affected by the prices farmers receive for their outputs and have to pay for their inputs. What looks like satisfactory irrigation system performance to the government, in terms of gross physical production, may be judged unacceptable to farmers, in terms of their resulting income.

What commodities will be produced on irrigated land may be a matter of contention between farmers and officials, as the latter may prefer and promote cash crops which farmers do not find profitable or which involve more risk or effort than farmers judge worthwhile. Or farmers may prefer high value crops, which require more water than approved by the government for that area (for example, in India, where cropping "zones" are established). Greater scope for farmer participation will tend to lead to production of crops which farmers prefer. This can be at variance with what agencies want to promote.

Improved water distribution is welcomed in principle by farmers as well as officials, tough farmers will usually emphasize reliable and predictable deliveries with upstream water users not necessarily wanting equitable distribution if it means reallocating water away from them. Governments endorse equity in water distribution but often find promoting it difficult, because their personnel may not be able or inclined to achieve it. We have been impressed in our literature review by how often farmers have devised methods organizational or technical -for equitable water distribution, suggesting a collective value placed on this objective. However, consensus is not always reached in specific instances where interests come into conflict and definitions of instances where interests come into conflict and definitions of "equity" can vary considerably from place to place, and among users.

Conflict reduction is often of special concern to farmers, though agencies also have an interest in promoting it, if only to diminish the problems it causes for them. it is hard to know how much value is attached to this objective because of difficulties in measurement. We have found in Sri Lanka that reduced conflict appears to be one o the payoffs which induces farmers to make special collective efforts to achieve a more equitable distribution of water, A low level of conflict:may mean either that there are few sources of strife in the situation or that existing institutions, formal and informal, are capable of controlling it. In the next chapter we will be considering conflict resolution as a generic activity invariable required for irrigation operation.

Governments often equate farmer participation with resource mobilization. Where an irrigation agency must "pay its own which the agency is most concerned. The innovative efforts of the National irrigation Administration in the Philippines to help farmers organize and to participate in design and construction activities were prompted in part by a policy decision to recover from users the capital in part by a policy decision to recover from users the capital costs of improvements. Farmers, for their part, may regard the possibility of mobilizing government resources as one of the reasons for them to organize. Farmers enthusiasm for resource mobilization will be conditioned by the extent to which the contributions required from them yield compensating benefits such as better water supply and distribution. To the extent that farmers contributions of knowledge are solicited in the planning or operation of systems, other contributions will be more readily forthcoming from them.

The goal of sustaining system performance should be generally agreed upon, but because it represents a deferred benefit, it is often overlooked. It is not clear whether farmers or officials tend to be more interested in this issue. There are several cases in our study which show declining production and capacity to support the population. Yields are reported as already declining in the relatively new Rahad scheme in the Sudan, while they have been stagnant for the last 20 years in the famous Gezira scheme in the same country. In some instances a decline is attributable to inadequate water control or neglect of drainage activities, resulting in waterlogging. It can also be due to a lack of incentives to invest in maintaining fertility. We expect that sustaining stable and productive irrigation systems is going to become a much more salient objective of system performance in the years ahead.

This last criterion highlights the possibility that efforts to increase production in the short-run can undermine capacity for long-term productivity. Installing too many tubewells, for instance, can draw down the water table over time, or excessive watering for short-run gains can result in salinization of the soil. Trying to achieve very high levels of resource mobilization from water users can similarly lead to conflict which reduces organizational capacity to achieve other goals.

It should be clear that there is not necessary symmetry between the goals of farmers and those of government agencies that engage in irrigation management. Such conflicts over objectives have been documented in a Philippine case by Svendsen. If government goals include attempts to maximize production while maintaining an artificially low price, for example, to keep consumers satisfied, the farmers whose crops are not very profitable will have little interest in "efficient" use of water which means more poorly remunerated work for them. Under such circumstances they will want to reduce weeding or to facilitate land preparation by irrigating "excessively" (in the opinion of officials).

Farmer participation will be more evidently useful and more sustainable to the extent that it is contributing to the achievement of objectives which farmers themselves value. Where the government seeks farmers' cooperation in activities whose outcome it values more than they do, it must be prepared to provide some compensating benefits which farmers appreciate, or to use some form of coercion. But such measures may not be successful, as seen in the Sabi Valley schemes in Southern Rhodesia and in the Rahad project in Sudan. Congruence of objectives between water users and system managers is among the most important features contributing to productivity as well as to farmer participation in jointly-managed systems.

Benefits of farmer participation

The literature gives only fragmentary evidence of precise gains from farmer participation in irrigation management, though it is replete with descriptions of benefits attributable to participation. Examples would include the report of a 30 percent increase in the flow of water to the downstream half of the Minipe scheme in Sri Lanka within the first year of introducing farmer organizations there. In the Pochampad scheme in India, the irrigable area was extended by 25 to 35 percent due to the rotational system which came into operation after Pipe Committees were established. With the help of these committees, the time required for land development of a turnout area (chak) was reduced from one year to only 4 to 6 months. In the Nong Wai scheme in Thailand, farmer organizations reportedly raised cropping intensity from 50 to 90 percent in two years' time. When a "participatory" approach was taken to expanding the Buhi-Lalo scheme in the Philippines, engineers with farmer advice and concurrence could reduce the planned length of field channels by 12 percent, thereby saving substantial costs. The construction work done by the farmers was completed four months ahead of schedule, and project engineers judged the quality of the work to be better than average.

Less precise but hardly less significant gains are reported from the Muda irrigation scheme in malaysia. When it was first opened, there was "anarchy", according to Afifuddin. Within several years this situation was replaced by some degree of order through the establishment of farmer organizations, which produced noticeable improvements in economic and social performance. In aggregated terms, Lowdermilk reports that farmers contributed $7.6 million worth of labor in a large program to rehabilitate turnout areas (chaks) in Pakistan. It is estimated that users are providing 30 percent of the cost for a World Bank rehabilitation project in Pakistan.

One can look at the benefits of farmer participation conversely, by considering the difficulties or costs where irrigation projects are operated without user inputs to management, even at lower levels of the system. The Gezira scheme in the Sudan was one of the first major agency-planned and operated systems, also one of the largest, over 1.8 million acres. Its early economic success encouraged other countries to embark on similar regimented, large-scale schemes such as the Mokwa project in Nigeria. Unfortunately, crop yields in Gezira have been stagnant for the last twenty years, and the case materials suggest that this is due more to social and organizational factors than to technical constraints. Tenants—as farmers in the Gezira scheme are called—do not see themselves as "partners" and thus have not been responsive to opportunities for innovation, according to the case documentation. To achieve more personal commitment and attachment to the project, its managers have now granted the Tenants' Union a voice in running the scheme, but the scope for farmer participation is still too limited to give them much incentive for change. Poor agricultural performance in the Egyptian case of Abu Raya was similarly attributed in part to the lack of self-respect and self-confidence possessed by farmers who have no active role in irrigation management. This suggests certain economic costs of not having farmer participation, but such costs are very difficult to measure.

One kind of farmer participation often overlooked and not encouraged is involving users in planning and operations. The case materials document that farmers have not only social skills for problem-solving but also valuable technical knowledge about acquiring and controlling water. In China, irrigation authorities in Meichuan accepted a "melons-on-the-vine" strategy after seeing a farmer-built system with many small reservoirs storing and controlling water along a main channel. With farmer help, they constructed 21 small reservoirs and over 6,000 ponds to add almost 30 million cubic meters of storage capacity to the 27 million cubic meter capacity of the main Meichuan reservoir.

In irrigation schemes in the bills of Nepal, tunnels have been constructed through portions of mountains where the channels installed along the sides of those mountains have been frequently damaged by landslides. This requires considerable skill in design as well as construction. In Kenya, Marakwet farmers have developed a furrow system taking water from the Rift Valley escarpment by ingenious channels to cultivable areas up to 9 miles away.

While such techniques may not apply to large-scale schemes, they testify to the empirical knowledge which farmers can have of hydrology and engineering. Moreover, since most of the major opportunities for large-scale irrigation development have been identified and developed, additional area is likely to be added in less obvious and less favorable circumstances, that require more inventive design and which could benefit from the low-cost construction techniques that farmers may know or be able to devise.

There is no guarantee that water users will always know or be able to apply appropriate technologies for irrigation. A case in point from our literature review is one where the Bangladesh Agriculture Development Corporation installed a deep tubewell at Pultan Para. It was up to the group of users to lay out and build a system of major and minor field channels, which they did rather badly. In such a situation, more technical advice from BADC would have been helpful for making the best use of available water. The extent of farmer competence to contribute to solving technical problems is something to be examined rather than assumed. presently it is too often assumed that users have nothing to contribute technically.

Costs of farmer participation

The Bangladesh example just discussed suggests the possible negative side of farmer participation. Also, we know that practically anything which has benefits is likely to have some associated costs, and participation is no exception. This suggests that participation should be "optimized" rather than maximized. Practically all kinds of participation, discussed in the next chapter, have some costs o farmers, if only in terms of time, whether representing forgone earnings or leisure. A participatory system of water distribution which enforces equitable division between head and tail areas may take away some of the advantages enjoyed by head-end farmers, which is a cost to those users. To agency mangers of irrigation systems, a greater farmer role may be seen as reducing their authority or status (or even income gained illicitly). Some of these costs may not be judged worth taking into account, but they will affect the incentives farmers and officials have for cooperating in a more participatory system of irrigation management.

In our review of the literature, we found many more comments on positive effects of participation than negative ones, though this may reflect some bias in the literature. Several cases were reported in the previous section where there were costs of not having farmer participation. We found that most of the time where changes were introduced by farmers or by agencies, it was in the direction of getting more rather than less user participation. Occasionally we found governments reducing the scope of farmer participation as done initially in the Sederhana project in Indonesia. In this case, when communal irrigation schemes were rehabilitated by an agency, it took over responsibilities that farmers had previously handled themselves. As this was costly and unnecessary, the government began experimenting with programs similar to those in the Philippines and Sri Lanka, discussed below, employing organizers to promote farmer participation in rehabilitating and then operating and maintaining the small-scale systems.

Even some schemes controlled by government agencies have moved, however haltingly, toward providing for more farmer participation. The Muda case in Malaysia and the Gezira project in Sudan were mentioned previously. In the Mwea scheme in kenya, agency managers have sought to introduce some organizational channels for farmers participation, even if only as "a useful safety value". Case materials suggest that a larger role for farmers would have been useful not just in the Tenant's Liaison Council and the Coop Credit Union but also in irrigation matters as well.

Comparisons of benefits and costs

Where farmer participation has developed spontaneously or is simply encouraged without any agency efforts to promote it, there are no costs or they are likely to be small. Agency personnel may feel that they are losing something in terms of status and power. But economical gains from the systems can offset this, and even give enough satisfaction and credit that staff accept this new arrangement. There may be cases where the efficiency of water use and/or the equity of water distribution is so satisfactory that it would decline if water users had greater responsibility for management. But such well-run systems are definitely a minority and probably have already developed satisfactory mechanisms for farmer involvement. Judgments must be made in each case about how close an irrigation system is to some optimum with respect to the criteria discussed at the beginning of this chapter, and whether performance levels could be increased through farmer participation.

Where irrigation systems are found wanting and a greater role for water users appears useful, the question becomes how to establish this. There have been a number of efforts made in various countries, discussed in the following chapters, to introduce farmer organization, commonly called water user associations, to achieve more farmer participation in irrigation management. It can reasonably be asked whether such efforts are cost-effective, whether the benefits therefrom exceed the costs.

This issue is difficult to answer definitely since there are many intangible benefits and costs that elude any summary measure comparing the two. Such factors should be taken into account, but policy-makers and planners usually want to know at least the ratio of benefits and costs that can be denominated in monetary terms to see what the narrowly economic impact of such a program would be. While such measure are likely to be only partial reflections of what is accomplished for a given expenditure, they should be estimated, accompanied by whatever qualifications and additional calculations seen warranted.

We have been able to find in the literature only two systematic comparisons of benefits and costs where farmer organization has been introduced to improve irrigation management. As seen above, there are various reports of benefits or costs, but seldom are the two compared for the same program. These two estimates, from the Philippines and Sri Lanka, are very encouraging, suggesting economic rates of return in the range of 50 percent. This is not considering intangible benefits, which probably exceed any unaccounted costs by an even greater margin. Such a rate of return on "software" is several times greater than that accepted now for investments in irrigation "hardware"

In an analysis of benefits and costs in 19 pilot irrigation scheme where the participatory approach was followed in the Philippines, direct quantifiable benefits (savings) in the construction phase were $24 per hectare, against a cost for the community organizer program (salaries,training, etc.) which came to $49 per hectare. This left a negative balance of $25 per hectare in direct costs at the time of completing rehabilitation. (Even this cost could be seen as offset by the fact that field ditches, which cost $38 per hectare to build, remained intact and were used, thanks to farmer consultation and involvement, whereas in many other schemes they were torn up within a few years; but this was not included within the benefit-cost assessment.) Since the value of farmer resources mobilized for operation and maintenance was calculated to be $12 per hectare annually, in narrow financial terms, the cost of the program could be "recovered" within some two years, and the stream of benefits should continue thereafter without additional or with little investment.

Where farmer participation was introduced in a large irrigation scheme in Sri Lanka, definite net benefits could be seen within two years. In a pilot area of over 10,000 acres, where organizers had been fielded to help farmers establish water user associations, the cost of the program including all training, supervision and salaries, was about 60 rupees per acre per season. Direct benefits from increased production came to about 90 rupees per acre per season, figuring only the value of maintenance work done by farmers and of increased production from just one tail-end area that remained uncultivated before farmers worked out and implemented a system of rotational water distribution. Now that the program has been established and can move into a "maintenance" phase with less intensive support from organizers, the ongoing cost including supervision, transportation, on-going training, etc., comes to about $1 per acre per season. This is more than justified if improved farmer operation and maintenance can raise production by even 1-2 bushels per season, an easy target to meet.

Additional benefits not include in Wijayaratna's calculations because of lack of data or difficulties in quantification, were (a) reduced damage to physical structures by farmers and animals, (b) reduced conflicts over eater, and (c) yield increases attributable to more reliable water distribution at the field level which encourage adoption of new technology. By calculating the marginal economic value of irrigation water, benefits could be attributed to increased efficiency in water use; alternatively one could value the production from the additional area that could be cultivated due to more sparing use of water upstream.

Still more difficult to measure is the value of improved system performance indicated by marked reductions in the number of irrigation-related complaints since farmer organizations were established. This represents a level of satisfaction that can translate into social and political benefits. Thanks to physical rehabilitation of the Gal Oya system, the facilities have been improved, as has the performance of the Irrigation Department, partly in response to being able to (or having to) work with organized water users. The main reason for fewer irrigation complaints being brought to senior politicians and officials has been the greater cooperation among farmers, who could solve many of the problems by themselves once organized, and cooperation between farmers and government staff, who have become more engaged in problem-solving with the result that higher levels are less often bothered.

The Philippines and Sri Lanka programs have encountered many difficulties and both have fallen short of their own goals in many ways, so they are not perfect "models" to be replicated. Many positive lessons can be learned from their experiences, however, and from the "learning process" approach adopted in both. Programs for introducing water user associations into schemes of irrigation management could be more successful or less successful than these.

The relevant consideration is that agencies bearing the cost of establishing organized farmer participation have realized significant returns from such investments, in the range of 50% according to these cases, considering only measurable and tangible benefits. This should encourage government and donor agencies to look seriously at the possibilities for trying to improve irrigation management by involving farmers more systematically.

9

What Kinds of Participation

Activities in irrigation Management

The interaction of physical and organizational aspects of irrigation makes it a socio-economic process, exemplified by the three focuses of irrigation activity which are closely linked with one another:

- Some activities focus on the water which is to be provided in an adequate and timely manner to crops;
- Other activities focus on the structures which give control over the water for its application to crops; and
- Still other activities maintain the organization of effort which can manage the structures that control the water.

It is probably not coincidental that these three focuses correspond to the three factors of production which economists classify as: (1) land, the term used for all natural resources; (2) capital, created from other resources to make them more productive; and (3) labour, covering all human activity. Water is a crucial natural resources, generally renewable within some limits. The physical structures for irrigation, like other kinds of capital, are produced through investments of materials and labour. Organization is established and maintained through human efforts, embodying both energy and ideas, which may come from users, from agency personnel, or from some combination of the two.

Since our concern here is with what users can do to improve irrigation management, on their own or as part of a more complex system of organization that includes agency staff, agency activities are considered mostly in relation to their support of effective user roles. The analysis in this chapter would apply with appropriate modifications similarly to purely agency-run schemes. The analytical framework for assessing farmer participation possibilities is presented first in summary form. Each of the sets of activities is then reviewed on the basis of what can be learned from the case materials.

The first set of activities focuses on water use:

1. *Acquisition* of water from surface or sub-surface sources, either by creating and operating physical structures like dams, weirs or wells, or by actions to obtain for users some share of an existing supply.

2. *Allocation* of water by assigning rights to users, thereby determining who shall have access to water.

3. *Distribution* of water brought from the sources among users at certain places, in certain amounts, and at certain times.

4. *Drainage* of water, where this is necessary to remove any excess supply.

These activities apply and must be dealt with at every level of a system, as analyzed in the following chapter.

Farmers may be active in any or all of these tasks, directly or through representatives, at any level.

Other activities deal with structures for water control. There is already a standardized classification for delineating such activities with regard to physical structures:

1. *Design* of structures such as dams or wells to acquire water, channels and gates to distribute it, and drains to remove it.
2. *Construction* of such structures to be able to acquire, distribute and remove water.
3. *Operation* of these structures to acquire, distribute and remove water according to some determined plan of allocation.
4. *Maintenance* of these structures in order to have continued and efficient acquisition, distribution and removal of water.

Each of these activities relates to and facilitates the preceding water use activities. They are as relevant to organizational structures as to physical ones. While the structures required for acquisition, distribution and drainage of water are basically physical, those for its allocation are essentially legal or contractual. A capacity for allocation needs to be planned, established, operated and maintained just as surely as does the capacity of a reservoir or a drainage system. Even if allocation activities are not as material as those for acquisition, distribution and drainage, the parallels in terms of the activities involved are substantial as seen in following table.

Relation between structure-focused and water-focused activities

Activities associated with water control

Water use activities	*Design*	*Construction*	*Operation*	*Maintenance*
Acquisition	Design of Acquisition Structures	Construction of Acquisition Structures	Operation of Acquisition Structures	Maintenance of Acquisition Structures
Allocation	Decision on water Allocation	Establishment of water Allocation System	Operation of Water Allocation System	Maintenance of Water Allocation System
Distribution	Design of Conveyance and Control Structures	Construction of Conveyance and control Structures	Operation of Conveyance and Control Structures	Maintenance of Conveyance and Control Structures
Drainage	Design of Drainage Structures	Construction of Drainage Structures	Operation of Drainage Structures	Maintenance of Drainage Structures

One sees from this how various irrigation activities are undertaken both with reference to a particular phase of water use—acquisition, allocation, distribution, or drainage—and to accomplish some kind of control over water in these different phases. The structures involved may be physical, legal or organizational, but all kinds of structures need some design or planning, some construction or implementation, some process of operation, and some activities of maintenance.

Going along with each of these activities which focus on water or control structures are certain organizational activities that marshall human efforts, to make collective action more predictable and effective. These activities can focus on the structures, on the resources of water, or on the irrigation organization itself.

The four basic organizational activities, already introduced are:

1. *Decision-making:* This applies to acquisition, allocation, distribution or drainage of water; to design, construction, operation or maintenance of structures; or to the organization which deals with these activities. *Planning* is one major form of decision-making.

2. *Resources mobilisation and management:* This involves the marshalling as well as application of funds, manpower, materials, information or any other inputs needed for the above activities, or for any general organizational tasks.

3. *Communication*: This concerns the needs and problems in any of the activity areas noted above, conveying information about decisions made, about resources mobilization, about conflicts to be resolved, etc. to farmers or any other persons involved in irrigation. One purpose of communication may be *coordination*.

4. *Conflict management*: This must deal with differences of interest that arise from activities of acquisition, allocation, distribution, drainage, design, construction, operation or maintenance, or from organizational activities generally.

One can have more or less farmer participation in general terms (e.g., how much user participation is there in decision-making or in communication?) or specifically (e.g., how much labour is being contributed in resource mobilization for maintenance? or in the conflict management associated with water distribution, who resolves disputes over water rotations?).

Organizational management activities refer both to physical objects like water or gates and to social relations among people. Resources mobilization deals mostly with material resources but also with non-material things like information an ideas. Even acquiring water through dam or pumping facilities is thoroughly socio-technical because decision-making, resources mobilization, communication, and conflict management are intimately associated with the physical structures and resources flows.

The four organizational management activities closely parallel the preceding set of activities aimed at gaining control over water through physical or social structures, as seen from the following comparison:

STRUCTURE ACTIVITIES	*MANAGEMENT ACTIVITIES*
DESIGN/PLANNING	DECISION-MAKING
CONSTRUCTION/IMPLE-MENTATION	RESOURCE MOBILIZATION
OPERATION	COMMUNICATION/ COORDINATION
MAINTENANCE	CONFLICT MANAGEMENT

This similarity does not make them identical, however, because the management activities on the right-hand side apply to each of the activities in the left-hand column:

Design obviously entails making decisions, but it also requires mobilizing information, having communication among the relevant actors, and resolving any conflicts that arise over the design itself.

Construction or implementation will involve many decisions about how something will be carried out, substantial resource mobilization, much communication and coordination, and reconciliation of divergent interests and opinions as the work is done.

Operation requires decision-making about schedules, work assignments, etc. the mobilization of resources like information, labour and funds, regular communication about schedules, resource contributions, etc., and conflict management to the extent there are any disagreements about operation.

Maintenance likewise calls for decision-making, much mobilization of resources, considerable communication, and also handling of disputes over what is to be maintained, how, any by whom.

From this we see how inter-related all irrigation activities are. They can be viewed from any one of the three perspectives. If one's analysis is water-focused, one still needs to look at the structures and the organized efforts that give control over water. if one focuses on the structures, these have to be assessed in terms of how they affect water flows and what organizational activities they require. Alternatively, if one takes an organizational perspective on irrigation management, these activities have little meaning except as directed toward the physical as well as social relationships subsumed.

Water use activities

Activities focused on water appear to be the most direct forms of irrigation management. This is often seen dramatically in the acquisition of water through design, construction, operation and maintenance of facilities such as weirs across rivers, bunds forming catchment reservoirs, or wells tapping underground sources of water. The water thus acquired for use needs to be allocated among uses and among users according to some plan or set of criteria. Water is not assigned to uses or users "naturally", so allocation is an essential irrigation activity, even though once a system of rights has been established, there may be nothing to observe. Allocation schemes when they get implemented through the distribution of water, on the other hand, are very visible.

A system of conveyance and control structures is needed to transmit and distribute water according to some plan. Like acquisition facilities, canals, channels, regulators, gates, etc. have to be designed, built, operated and maintained. Whether or not drainage is an explicit activity in an irrigation system depends on such physical features as topography, soil type and climate. Drainage is frequently neglected as a part of irrigation, but removal of water where this does not occur naturally is crucial for sustained system performance.

What are appropriate or even necessary water use activities for farmers will depend on the nature of the system, and the level at which users as well as agency personnel have responsibility. We analyze the levels of operation and organization within a system, and in the subsequent chapter, the various roles which farmers and professional can perform in irrigation management. Making distinctions of where water use activities occur and who undertakes them will refine the analysis. But consideration of farmer participation in water use activities can proceed here with these other dimensions of analysis having been noted as relevant.

In a simple one-level irrigation system, depending on a run-of-the-river diversion barrage, a small catchment dam or a deep tubewell, and serving perhaps several dozen acres, activities of acquisition are usually carried our entirely by the users. In contrast, within huge systems such as those in India, Pakistan, Egypt or Sudan, farmers'

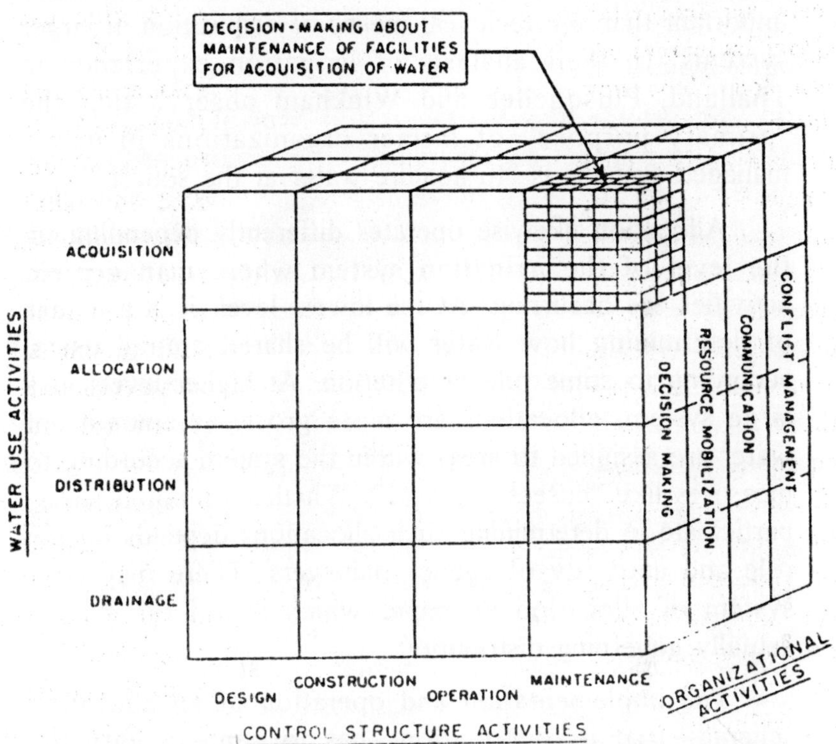

Matrix of irrigation management activities

involvement in acquisition is quite different. It may involve lobbying with the agency that controls the water to get both allocation and distribution of water to a particular part of the command areas. Acquisition may require farmers to protect the water allocated to them to make sure it reaches their channels, as water is sometimes acquired by stealing what has been allocated to others. Acquisition thus can involve quite different activities depending on the size and complexity of the irrigation system and on the level at the activities occur. When focusing attention on farmer activities in irrigation management "above the outlet," Chambers find that organizational strategies for water acquisition are more important than are technical means of acquisition in larger systems. In their analysis of irrigation experience in Thailand, Plusquellec and Wickham observe that the "primary purpose" of farmer organizations in user-managed systems is "to acquire water at the source."

Allocation likewise operates differently depending on the level of the irrigation system where management activities are occurring. At the lowest level, it is a matter of determining how water will be shared among users, according to some rule or criterion. At higher levels in a large system, allocations are more gross, as amounts of water are assigned to areas within the system according to measures like "cusecs per acre". Whether or not farmers participate in determining such allocations depends on the role and authority of agency managers. There may be a system of allocation on paper which is not the system actually governing distribution.

The implementation and operation of an allocation scheme—that is, distribution—has to contend with the universal problem of locational advantage, where upstream users have greater opportunity to obtain their share (or more than their share) than do users downstream. In the absence of organization (or of powerful users downstream), inequalities in distribution are frequently observed, and organizations are not uniformly successful in promoting equitable allocation and distribution. One of the main purposes of having water user associations is often to deal with this problem.

Because people's livelihoods may be at stake when water is very scarce, distribution becomes more difficult under such circumstances. Indeed, they may be two different schemes of allocation and two different regimes of water distribution between the wet season and dry season, as reported in the Philippines, Indonesia, Mexico.

Farmers' ability to handle distribution when water is scarce depends on numerous factors. In many countries, users have devised ingenious means, both technical and social, to distribute water equitably. Examples would be the proportioning weirs found in the hill systems of Nepal and the subaks of Indonesia, and the traditional practice in some schemes in Sri Lanka and the Philippines of locating at the tail-end of the command area the plot of land which is given as compensation to the person responsible for distributing water.

There are situations where farmers would like to have someone with authority from outside the community handle the tasks of distribution, as a way of reducing conflicts or avoiding laborious efforts. Lowdermilk et al. report that Pakistani farmers were glad to have an outside agency distribute water among groups, according to strict rules. However, they preferred to handle water distribution within their groups themselves, modifying the official warabandi rotation by making ad hoc informal arrangements for water-trading (which the agency had declared to be illegal) in order to better meet farmers' individual needs.

Users should be able to distribute water among themselves, though this does not mean they will always handle this task, as there may be competing demands for their labour or social conflicts may interface. In Taiwan, where farmer responsibility has been taken to the highest organized levels of any country, it is reported that the irrigation bureaucracy now sometimes employs temporary labourers to distribute water at the field channel level oo this themselves. One should not generalize for all levels or all countries or all times about what users can and will do.

We found little reference in the case studies to farmer involvement in drainage activities. This probably reflects one of the "blind sports" in the literature, as discussions of irrigation commonly neglect drainage. Improving water control among users can help to alleviate drainage problems so this activity should not be viewed in isolation. Any re-use of drainage water, something farmers are often better able to plan and manage than are engineers, can often increase irrigation efficiency. Certainly where condition require drainage activities, they must be performed by someone. So an analytical framework should include consideration of drainage, whether or not it presents problems for users and/or agency personnel to deal with.

Control structure activities

The design of structures to control water from a source (acquisition) to farmers' fields (distribution) and beyond (drainage), according to some guiding structure of rules and procedures (allocation), can be done by users or by various specialists. Generally the larger and the more complicated the system, the more the latter are needed to introduce scientific principles and technical information into the design. However, this does not mean that farmer participation in design activities should be limited to small schemes.

In the Quinua study from Peru, one finds impressive ways in which irrigation systems have been designed to serve land in a vertical series of environmental zones, each with different conditions. A limited amount of water is distributed in a parsimonious way at successive times to different places at different altitudes to be used for different purpose "in a most economical dovetailing of functions". The largest system in our sample designed without agency involvement, Chhatis Mauja, was established 150 years ago in the plains of Nepal and covers 7,500 acres. It presented some difficult technical problems because of the rapid flow of the Tinau river during the monsoon season, but the users' design of an organization for operation and maintenance is probably even more impressive than the engineering aspects of their system.

Construction is a task frequently undertaken by users, often under the direction of technical personnel. The ingenuity of the Marakwet in constructing their schemes in Kenya—including suspending channels 15 feet above the ground— was described in previous chapter. Of course, the technical activity of construction is not possible without the organizational activities involved in resources mobilization that make construction possible. Users often modify the structures in systems designed and constructed entirely by agencies, which constitutes re-design and reconstruction. This may or may not improve system performance.

Operation and maintenance are generally treated together, perhaps because they so often occur concurrently, though they are in fact quite separate activities. We will be discussing farmer participation in operation and maintenance when considering irrigation management at different levels in the next chapter, and in relation to specialized roles. It suffices here to highlight again the fact that maintenance of control structures should be distinguished from the management of the water within the structures. Fleuret observed in his study of irrigation in the Taita Hills of Kenya that, "For the most part, different social groups undertake the two tasks (management of structures and management of water) in different ways at different times".

Organizational activities

Individuals make decisions and they mobilize and manage resources, but there are limits on what can be accomplished without collective action. By their very nature, communication and conflict management require involvement of more than one person. In irrigation management, the activities of decision-making, resources mobilization and management, communication, and conflict resolution encompass the main focuses of common effort among users. Moreover, they represent also the ways in which users can participate in irrigation management at higher levels within the system, as will be discussed in the concluding section of this chapter.

Decision-making involves more than simply deciding on a course of action. It requires evaluation of performance, identification of problems, gathering information, formulating alternative, solutions, building consensus, with decision-making consolidating the various activities. This process can be:

— formal or informal,

— routinized or ad hoc,

— undertaken by all the persons affected or by their representatives,

— binding on all concerned or advisory only.

The decisions taken can vary in terms of their being:

— routine or innovative,

— major or minor (according to whether substantial or small commitments of resources are involved), and

— independent of other decisions or contingent on them.

The kind of decision-making structure that exists does not necessarily correlate with the size and complexity of the system. The Sonjo in Tanzania have a very elaborate "traditional" structure of decision-making for managing their small-scale systems. They have four categories of system membership, but only one category, the major elders, has any decision-making authority. In the "modern" Mwea system in Kenya, and elaborate structure has been established with a Tenants' Liaison Council and Tenants' Advisory Committees, but still there is little user control over water decisions. These are handled through the line of administrative authority which descends from the National Irrigation Board.

The frequency and kind of decision-making needed can vary among systems. *Qanat* systems found in Iran, Oman and other countries of the Middle East and North Africa are said to run practically "on automatic", with islamic precepts guiding users' behavior. Village authorities become involved with irrigation matters when such systems require repairs, but little O&M activity is needed as the systems are mostly underground, operating according to designed capacities and receiving practically no maintenance. The significant decisions for such systems are the initial ones concerning investment and construction or infrequent decisions about rehabilitation.

The *ahar* systems in Bihar state of India require more

management effort but decision-making is handled informally by local elites. These systems are created by a network of catchment reservoirs connected by distribution-drainage canals (*pynes*). Prominent farmers whose land is served by reservoirs at the head of the system are expected to mobilize labor periodically from all the areas served to do maintenance work on the *pynes*. They organize and oversee communal labor whenever it is needed. These examples of minimal decision-making, however, indicate that there is always at least some decision-making activity. The question is, *who* will do it, and *how?*

Resource mobilization is the most visible organizational activity in irrigation management, directed most dramatically towards construction as a one-time effort or more commonly to maintenance as an ongoing activity or for rehabilitation. Labor is the resource most extensively mobilized, though money and materials are also important; likewise farmers' information should be regarded as a major available resources. One of the best examples of possibilities for resource mobilization is the Nepal system mentioned above, Chhatis Mauja. Its 4,000 farmers contribute 60,000 man-days of work annually for desiltation and maintenance of their main canal. In addition, area and village committees organize operation and maintenance activities within their respective jurisdictions. The smaller systems of Argali and Chherlung in the hills of Nepal mobilize 1,500 to 2,500 days of labor each year (10 to 30 man-days per acre), and both organizations have also raised cash from members to line their canals with cement.

Lesser contributions of resources may be needed in other systems less vulnerable to damage and less beset by siltation. Farmers at San Pedro de Atacama in Chile contributed each month one man-day per hectare for main canal maintenance while also paying a fee of 2 pesos per hectare to cover the ditch tenders' fees. Substantial labor mobilization is similarly reported for traditional schemes in Peru and Mexico.

Even an impressive amount of resources mobilization will not produce the expected benefits unless there is good resource management. This can be provided by users in many circumstances. For example, in the Seraphi system in Thailand, labor responsibilities are assessed quite precisely according to area served, and careful records are kept to ensure that all make their contribution. Very precise and skillful resources management is also documented for the *Zanjera* systems in the northern Philippines. As a spur to performance, for example, thee is competition among work groups assigned to rebuild respective sections of the weir, to see which can do the best job. Management tasks may be handled through specialized roles like "irrigation headmen" or through other institutions. Among the El Shabana in the huge Daghara system in Iraq, contributions of labor and cash for irrigation operation and maintenance are collected through the structure of "tribal" organization rather than through explicit irrigation associations.

Whether it is easier for users to contribute labor than cash will depend on their circumstances. Where money incomes are low, farmers usually prefer providing labor, but when there are good opportunities for wage employment or other claims on farmers' time, they may wish to make payments instead of participating in work parties. Some form of labor mobilization is common in most systems, at least to deal with O&M requirements at lower levels. Still, there are limits to how much of the costs of irrigation management can be covered by labor contributions.

Some interesting "hybrid" systems of resources mobilization can occur. In the Nam Tan system in Laos, farmers paid, the "traditional" irrigation headman 16 kilograms of rice for every hectare covered by his services and paid the irrigation agency an additional 80 kilograms of rice for every hectare for its expenses. This amounted to 5 per cent of average output. Similarly, the Dharma Tirta irrigation organizations in Indonesia combine contributions of labor and cash, with the fee paid at harvest time and in relation to farmers' yields. Some portion of the funds raised is set aside for further investment, and those organizations which undertake major investments to improve their systems can win prizes from the government for their accomplishment.

There is often resistance reported to requirements of cash payment from users. But Singh reports that 80-85 per cent of farmers in the large Pochampad system in Andhra Pradesh, India were agreeable to a charge provided that the money would be used to improve their service and would not be misused. The O&M requirements of the system were such that some of the resources mobilization needs could be met with labor for maintenance work below the pipe outlet. With 2,000 water user associations (Pipe Committees) in the scheme, Singh estimated that work worth $1.6 million annually could be covered by farmers.

One ingenious method for mobilizing cash directly to meet certain organizational needs is reported in the Izki system in Oman, where water is allocated not according to land area but according to "shares" Although most of these are owned by families, the organization holds some shares which are auctioned on a weekly or annual basis to meet expenses. Capital was mobilized for expansion of the Chherlung system in Nepal by selling off water "shares" in a similar manner.

Mobilizing resources can be facilitated by having a clear and acceptable division of responsibilities between water users and the agency. A good example is the arrangement worked out for pump irrigation in the Senegal river basin, where water users cover the costs of land development, cropping, pump operation and maintenance by contributing labor and money. The government agency provides technical assistance and supervision, while a donor agency contributed the capital costs of the pumps. Work groups of users appear to be handling their part of the resource mobilization quite satisfactorily, partly because all internal responsibilities of organization and conflict management are left to the group. Without such a clear division of responsibility, farmers and agencies may both leave certain tasks for the other to do.

Resources mobilization from users appears likely to be more successful when decisions on means and shares are left up to the users themselves. Lees in her study of 24 communities practicing irrigation in the Oaxaca state of Mexico found a great diversity of methods. For example, only nine communities levied direct charges on farmers for water according to their land area or the time water was received; others had more complicated systems which took into account people's contributions of time to village activities, etc. The two systems in Thailand studied by

Abha had a number of different means for mobilizing resources to cover organizational and investment costs:

(1) cash contributions

(2) grain contributions,

(3) exempting certain persons who had organizational responsibilities from their labor obligations,

(4) fines,

(5) selling shares of water, and

(6) selling land, or conferring usufruct rights.

A standardized system would be less able to tap sources that were accessible and plentiful and from which people were most willing to contribute. Flexibility and diversification are important features for any resources mobilization scheme. Furthermore, scheme for resources mobilization are likely to be more sustainable to the extent that there is honest and efficient resource management.

Communication is an organizational activity "so universal that one doesn't see it" (Robert Chambers, personal communication). Its purpose is to help with coordination, which is vital to the discharge of all other irrigation functions. One of the tangible evidence of this functions is the creation of specialized roles, discussed, to handle communication in a number of systems. The Marakwet in Kenya, for example, have set up a system of communication whereby the bad news of need to mobilize labor for a major repair is relayed to all concerned by appointed "blowers", persons who reside at selected locations and are given special horns. The central committee operating the Chhatis Mauja system in Nepal includes two appointed "messengers" who communicate with the 54 member villages about general meetings and dates for maintenance work. They are given a small cash payment plus grain and the use of a bicycle. Similarly in Thailand, in the Seraphi scheme, assistant irrigation headmen serve as "runners" to transmit messages from the headmen to farmers and also coordinate irrigation activities in sections of the village assigned to them.

Failures in communication can exact costs in terms of system operation and maintenance. Studies in irrigation systems in India and Pakistan have found that between 70 and 83 per cent of farmers did not know the dates when they were expected to do maintenance and repair work or even the dates when water issues would end. In these cases, even on-way communication—to farmers—was inadequate, and two-way communication which would convey farmer needs and capabilities to system managers was still more deficient. In the absence of organization among farmers, communication among them will be limited, reducing their possibilities for cooperation to utilize available water to best total advantage.

Conflict resolution is difficult to judge because where there is much observable "success", it may be because indirect or tacit efforts to avert conflict did not succeed. Where there is little or no strife, rules and procedures may have been devised that handle problems and disagreements so smoothly conflicting interests are adjusted before they lead to disputes or to blows. Or it may mean there was no clash of interests. In certain situations it appears that the need to cooperate for irrigation can overcome propensities for conflict that exist in the community.

Some communities and some cultures appear to have a disposition for conflict. This is suggested in several village studies from Pakistan. In the Daudzai case, village elders are called upon fairly frequently to settle even armed conflicts. In one portion of the Seraphi system in Thailand, conflict became so severe that farmers stopped cooperating and part of that system went out of operation. An aqueduct system of irrigation in one Papua-New Guinea community required the cooperation of two village wards for its maintenance. Periodically, conflict between the two wards became so great that the aqueduct fell into disrepair and the system of agriculture reverted to separate smaller irrigation systems, until cooperation could be resurrected to reconstruct and maintain it for a time.

In certain societies, on the other hand, there appears to be some aversion to conflict. Farmers in Abu Raya, Egypt are reported to want to avoid conflicts within their communities, though this did not rule out conflicts between communities. There was also little conflict within villages in the Oaxaca state of Mexico according to Lees, but disputes were observed between villages, usually over land rather than over water. Similar efforts to maintain good relations among people within a community are said to keep the level of conflict low in Daghara, Iraq, where people in irrigation communities are all from the same tribe. Conflicts which do arise are mediated by community members who claim descent from the Prophet Mohammed. There is strong conflict between tribal groups at Izki in Oman, but all depend so much on the *Qanat* that serves them that they cooperate to keep the system working—though it is reported that twice, fighting between two Izki groups almost wiped out the *Qanat*.

When there are absolute shortages of water, the likelihood of conflict usually goes up. In the Diaz Ordaz system in Mexico, when water becomes quite scarce during late October, conflicts become quite intense. The system of allocation is accordingly changed between seasons to take into account the different degrees of water stress on the organization. The complicated system in Quinua, Peru, described above, was able to operate in the past with very small quantities of water and managed conflicts reasonably well through a traditional hierarchy of "civil-religious" roles. After these roles were abolished by law in 1970, however, the irrigation system became "acephalous" and was prone to much conflict. The legally-recognized "modern" municipal officials have been unable to govern water use, and the strongest individuals and groups are now able to dictate water distribution.

Conflict management may work better through informal mechanisms, such as provided by "traditional" roles and institutions, than through legalistic ones. So long as the authority of the tribal elders among the Marakwet in Kenya and the Sonjo in Tanzania remained intact, conflicts have been managed with little difficulty. The traditional irrigation headman role in Sri Lanka, the *vel vidane* role documented by Leach, was abolished in 1958 and his responsibilities were vested in elected bodies of water users (Cultivation Committees). These were in turn abolished in 1977 and replaced by an appointed Cultivation Officer . Not only conflict management but most other irrigation activities have suffered, with the result that the government is now seeking to establish water user organizations building on old and new "traditions". The key element for effective conflict management roles is not whether they are "traditional" but whether they enjoy the confidence of water users. Imposed roles from outside are not likely to have this, though roles evolved with users' knowledge and cooperation could.

Conflict resolution as an organizational activity resembles drainage as a water use activity. Because it may not always be necessary, it is taken for granted more easily and more often than other activities. Moreover, the preferred situation is where conflicts, like removing excess water, are handled gradually, naturally and imperceptibly.

External organizational activities

Organizational activities have been discussed thus far as "internal" to a particular set of water users, who make decisions about what they should do collectively, who mobilize resources from members, communicate and resolve conflicts among themselves. In fact, each of these activities can be undertaken 'externally" with reference to other water users or with officials who operate at higher levels. We will not explore this distinction at length here because it requires consideration of levels of organization, the subject of the next chapter. It should be clear that organizations for irrigation management operating at one level, such as the field channel or distributary, may participate in management tasks at other levels.

Farmers managing a distributary canal, for example, may be involved through their representatives in decision-making that allocates water among canals, something which these particular farmers cannot decide on their own. When facing a major repair problem, users may mobilize resources from a government agency—heavy equipment or subsidized cement purchases, for example. Some of the communication which an organization undertakes will be with persons or agencies outside its own ranks, and a good share of conflict management activity will involve negotiations with other groups. A number of the cases cited above suggested that conflict involving water is often more serious between communities than within them.

The four organizational activities identified in our analysis thus apply to water management relations both among users at a particular level and with users and officials at other levels. To understand better the possibilities and problems of farmers organization for improved irrigation management, we need to consider the matter of "levels," to identify what kinds of participation can usefully occur where within irrigation systems.

10
Irrigation Legislation and Participatory Management

Historical Perspective

The Colonial Era: When the colonial government recognized its responsibility for irrigation development in the country, its approach evinced a sense of moderate cautiousness with a mix paternalism, humanitarianism, and self-interest. Colonial authorities were already aware of the adverse impact created by the Colebrooke-Cameron reforms of 1832 by which the ancient institutions of compulsory labour (rajakariya) and hereditary headmanship were abolished. Irrigation was one of the principal sectors affected by the reforms. The implementation of irrigation programmes therefore had to be undertaken with great care. The strategy was initially to resuscitate the ancient customs, traditions, and practices in the paddy secotr. For this purpose the Paddy Lands Irrigation Ordinance No.9 of 1856 was enacted for a limited period of 5 years. The justification for the proposed course of action is clearly stated in the preamble to the Ordinance as follows:

The non-observance of many ancient and highly beneficial customs connected with the irrigation and cultivation of paddy lands as well as the difficulties, delays, and expenses attending the settlement of differences and disputes among the cultivators relating to water rights, in the ordinary course of law, are found to be productive of great injury to the general body of proprietors of such lands and it is expedient to provide a remedy for these evils.

Restricting the validity of the Ordinance to 5 years presupposed that the careful monitoring of the implementation process would necessitate revisions and modifications. This illustrates an early perception by the colonial authorities of what is today called "a learning process".

Ordinance of 1856 entrusted the Government's responsibility for irrigation development to the Government Agent who was the administrative head of the Province. The GA was expected to perform his functions with the advice of the proprietors of the irrigated lands. In that role, the GA was deemed to function as a benevolent judge, implementor, and facilitator.

The same ordinance provides for the revival of the Village Council for conflict resolution in the course of implementing the law. The GA was required to preside in both meetings, the proprietors' meeting to obtain advice, and the Village Council meeting to resolve conflicts.

The implementation of irrigation programmes was constrained by the lack of funds from the central government. Government had no desire to increase its financial burden by recruiting village level functionaries. Therefore it was clear that reciprocal contributions by the beneficiaries should be the guiding principle to mobilize local resources in support of the programme.

The 1856 ordinance was revised by the Ordinance No.21 of 1867. In addition to the Village Council it provided for the selection of one or more headman by the proprietors to ensure the maintenance of rights and the prevention of any act militating against ancient customs and causing damage. However, the headman selected by the proprietors was made accountable to the GA. The same Ordinance demonstrated a remarkable degree of flexibility and understanding by allowing the proprietors to decide whether the operation and enforcement of the provisions in the Ordinance should be carried out with the aid of the Headman, the Village Council, or both.

After the enactment of the first ordinance in 1856 there were a score of amendments and revisions over the next 125 years. Since the irrigation Ordinance was expected to spell out the basis of organisation for irrigated agriculture, it throws some light on policies and perceptions that existed during the respective periods.

During the first two quarters of implementing the Irrigation Ordinance, a desire to monitor the implementation of its legal provisions was quite evident. Regional differences in irrigation practices were also recognized. The basic institutional framework enunciated that proprietors in an irrigation area should be allowed to decide for themselves the most desirable course of action, subject to certain limits of approval which do not seem to have impeded participation by farmers. An important feature in the monitoring process was that the colonial authorities relied on empirical evidence to support changes.

With the establishment of the Irrigation Department in 1900, some of the functions handled locally by the GA were transferred to the Director of Irrigation. Leonard Wolf, who held the post of AGA in Hambantota at the time, recorded his resentment in a diary. In his opinion framing cultivation rules was better done by the GA as an administrative function. However, after some time the status quo was restored.

Changes in policy perspective relating to irrigation development began to emerge in 1930s with an emphasis on the restoration of major irrigation works that lay abandoned in the dry zone parts of the country. With the eradication of malaria, prospects for the colonization of the dry zone and its irrigation development appeared to be brighter. By this time the ID had also collected adequate data on rainfall, streamflow observations, flood records, etc., and developed an expertise to handle major construction work. So the stage was set for a major transformation in irrigation development.

In the meantime local demand and pressure to improve existing irrigation works, largely, village works, continued. Provincial administrators were confident about the programmes under implementation. In the ID however, officials were reluctant to assign technical officers to what they called the excessive involvement with village works. It was argued that from a food production standpoint the village works were worthless as compared to the major irrigation schemes.

With the emergence of major constructions as the principal area of work by the ID, the role of the GA in provincial development grew even more important. The Government looked to the GA to coordinate and manage the resettlement of people selected under irrigation schemes opened up in the dry zone. The Land

Development Ordinance under which land redistribution programme was set in motion, conferred a special place for the GA to implement the colonization programmes. This was in addition to the functional roles already assigned to the GA under the Irrigation Ordinance.

The setting up of the District Agricultural Committee in the mid-thirties facilitated the GA's work as the principal coordinator of the irrigation programme in the province. This Committee, consisting only of officials, was incorporated into the Irrigation Ordinance No, 32 of 1946 to provide a legal backing to the decisions of the Committee.

A significant outcome of this change in perspectives in irrigation development was the enhancement of the decision-making power of the bureaucracy by a gradual process of imposing limits on participation by the farmer community. It is not clear however whether the new direction was the result of problems arising out of new dimensions in organisational management applicable to major systems. The protected tenurial system prescribed in the Land Development Ordinance, under which the newly reclaimed lands in the colonization schemes were distributed, required continuous supervision by officials. This may have had an impact on the irrigation management aspects too.

In the 1930s the emphasis was on resettling as many settlers as possible to achieve targets set by the policy makers. System design, especially in the tertiary levels, and the institutional framework for farmer participation, both of which evolved in the village works, were superimposed on the major system. The cultivation meeting is one such element, found to be ineffective in major irrigation systems with large number of farmers. However Irrigation Ordinance No.45 of 1917, section 18, provides for the proprietors to "appoint a committee of such members as they may determine to frame rules on their behalf, subject to confirmation at a subsequent meeting." The extent to which such a Committee was effective is not clear. It has been allowed to remain in the Irrigation Ordinance for about 50 years.

Post-Independence Era: In the period following the granting of Independence in 1948, four key issues, farmer participation, irrigation headman, conflict resolution, and maintenance, were dealt with by introducing amendments to the Irrigation ordinance.

Unlike in the earlier era, a noticeable tendency emerged to introduce conceptual changes in conformity with the official perceptions. Such changes were drawn more from abstract notions of a centralized system administration than from an empirical process of monitoring and evaluations. During more recent years, constitutional guarantees figure more prominently and seem to restrict the application of legal provisions embodied in the irrigation Ordinance.

Farmer participation in Irrigation management

The Irrigation Ordinance No. 9 of 1856 envisaged farmer participation at a public meeting of proprietors summoned by the GA. This was the embryonic form of the present cultivation (kanna) meeting. As the area under irrigation facilities expanded and the GA was unable to hold as many meetings as required, Irrigation Ordinance No. 16 of 1906 provided for the setting up of a District Committee of not more than 12 nor less than 3 persons to advise the GA on drawing up rules regarding cultivation practices.

In addition to the above District Advisory Committee, powers were given to the whole body of proprietors under Section 1 of the irrigation Ordinance No. 45 of 1917. The body of proprietors was to meet under the chairmanship of the GA to make rules on matters pertaining to the management aspects specific to each scheme which included the enforcement of ancient customs, irrigation headman, mobilizing farmer contribution, and system maintenance.

Furthermore, proprietors were empowered to meet under the GA and decide on the variations to irrigation rates, and to validate any irregularity, correct any informality, decide on matters referred to the proprietors by the Governor, and decide on bethma cultivation. A more significant feature in these provisions was that the proprietors were allowed to"appoint a committee of such number as they may determine, to frame rules on their behalf, subject to confirmation at a subsequent meeting."

With the enactment of the Paddy Lands Act of 1958, amendments to the irrigation ordinance became necessary. In introducing the amendments in Parliament, the Minister noted:

> Government Agents under the Irrigation Ordinance were more or less independent authorities...We find that there should be more control of the functions of Government Agents and closer coordination among them on the paddy cultivation side..... On the cultivation side the Commissioner of Agrarian Services is proposed to be brought in, an under his general direction and control the GA will work.

In fact, central control over the management of irrigation

systems was progressively increasing with the Government taking more and more interest in major irrigation systems. The 1968 amendment to the Irrigation Ordinance justified such increased control because the Government transformed major irrigation systems to food production centers. The exercise enjoyed only a short lived success. In a way, the new advances in agricultural technology also resulted in some alienation of farmers from the decision-making process due to the short-sighted policies adopted in implementing the food production programme.

A significant change in the composition of the cultivation meeting was effected by the Paddy Lands Act of 1958 which sought to introduce far-reaching tenancy reforms in the paddy sector. Tenurial arrangements of most lands in the government-initiated major irrigation systems are governed by the Land Development Ordinance. Therefore they are subject to a strict tenancy reform. But an amendment to the Irrigation Ordinance was introduced in 1968 to bring it in line with the Paddy Lands Act. The Irrigation Headman was removed and the Committee appointed by the proprietors was abrogated and both were replaced by the Cultivation Committee, which was the grass-root organisation envisaged by the Paddy Lands Act.

The tenancy reforms and the success in food production, the latter achieved through the lateral spread of Green Revolution technology, opened up new horizons for institutional development in the agricultural sector. For want of dynamism in the irrigation sector to diversify its attention from design and construction work, this opportunity was not seized upon. Reforms initiated by the agricultural sector were allowed to fill the gap, irrespective of their relevance and applicability to the irrigation sector.

The experience gained during the last 5 years has shown that institutional reforms and structural changes promoted by one sector without reference to the other sector sometimes result in a negative and adverse impact at the field levels. The lack of integration between the agriculture and irrigation sectors in policy formulation has been a major contributory factor to this situation. With the creation of new specialized agencies such as the Agrarian Services Department in 1958 and many others thereafter, the diagnosis of field level problems affecting farmers was marred by individual professional biases and divided loyalties. Even more important is the fact that farmer organisations came to be treated as a terminal facility available to the bureaucy with which to operate their programmes. This is one of the main reasons which constrained the continuance of these organisations at the field level.

The past experiences, have made us doubt the relevance of the cultivation meeting as a suitable forum for farmers. Under the INMAS programme, the three-tier organisation ranging vertically from bottom-level field channel organisation to the Distributory Channel Organisation (Sub-Committee level) to the Project Committee reinforces the decision-making process of farmers. In the absence of any other forum for all farmers to meet at least once during a season, it is desirable to retain the cultivation meeting as a mechanism through which recommendations made by farmer representatives and officials at the Project Committee level could be adopted for implementation in the entire project. Similarly, the cultivation meeting can provide an opportunity to farmers to articulate their views more openly and even represent minority viewpoints.

In this respect, it becomes necessary to redefine the status of the Agrarian Services Committee (ASC) provided for under the Agrarian Services Act in relation to the three-tier organisation emerging in major irrigation systems.

ASCs are the successors to the Agricultural Productivity Committees which were set up earlier above the Cultivation Committee with certain new functions. In the recent experimental programmes in farmer organisation in Minipe and Gal Oya; water users were promoted to erect a new institutional structure which is based on water as the key input. In effect the three-tier structure is an outcome of that effort. As a result, the role of the ASC is now confined to that of coordinating the supply of input services and related matters. Accordingly, the Cultivation Committee, as the representative body of farmers incorporated in the Irrigation Ordinance, has to be replaced by the three-tier project organisational framework.

Irrigation Headman

irrigation Ordinance of 1867, for the first time, provide the selection of one or more headman to carry out matters agreed upon by the proprietors. The headman was selected by the farmers but worked under the control and direction of the GA.

The Paddy Lands Act of 1958 removed the Irrigation Headman and replaced him with the Cultivation Committee. The Cultivation Committee was a creation of the tenancy reforms. The extent to which the removal of the Irrigation Headman is relevant to the Act's principal objectives can be explained only in the context of the overall socio-political environment within which the new Government of 1956 was brought into power. The removal of the Headman from the Irrigation Ordinance was completed by the 1968 amendment.

After a period of 20 years, the Irrigation Headman (Vel Vidane) was expected to reappear through the Agrarian Services Act in the form of a representative elected by the farmers in a tract. But the functions assigned to him under the Agrarian Services Act do not necessarily justify the attempt to make a Vel Vidane out of the tract representative.

Recent experiences in Gal Oya and elsewhere have shown that farmers themselves are not clear about the functions expected of the tract representative, especially in the major irrigation systems. The new group of farmer representatives thrown up by a process of facilitation in Gal Oya was found to be more acceptable to farmers. But it is not possible to remove the existing representative formally appointed under the Agrarian Services Act. It is now accepted that where such conflicts occur, the approach should be more conciliatory and endeavours should be made to evolve interlocking arrangements so that community responses would settle the differences in favour of the most feasible organisational arrangement.

In the three-tier organisational framework envisaged for major projects, the need to demarcate the area of authority in terms of hydrological boundaries, especially for the field channel organisation and the D-channel organisation, is now accepted. The DCO will remain the formal organisation which will federate representatives from field-channel organisations. It is therefore necessary to ensure that the Vel Vidane should come from this organisation as a representatives of farmers to carry out matters concerned with water allocation, distribution, and maintenance so that he would be able to function in his original role more effectively. It is also necessary to ensure that the appointment, remuneration, and dismissal of the Vel Vidane should be left entirely in the hands of farmers in the DCO with no accountability to any position in the bureaucracy. The Irrigation Ordinance should therefore be suitably amended to bring back the Vel Vidane in the above manner. As far as matters dealing with input coordination are concerned, the DCO may be requested to appoint another person as its representative to deal with such matters, leaving the Vel Vidane to deal only with matters concerning water.

Conflict Resolution

The system of Village Councils reintroduced through the first Irrigation Ordinance was directed towards compromise and not punitive action. This conciliatory approach ideally suited the purpose of conflict resolution in irrigation matters. Dispensation of justice in a VC, which was presided over by the GA, was facilitated by the creation of the Irrigation Headman. Farmers were given the option to decide whether they should enlist the services of the VC, the Headman or both. This feature is important because it recognizes the urgency and diversity of issues and circumstances under which rapid interventions has to be provided to sustain the integrity of the physical system and the efficiency of the institutional mechanism.

The character of the VC was changed by the enactment of the Village Communities Ordinance No. 26

of 1871 which dealt with matters of a broad nature more relevant to local administration. It was largely a creation of the officials with little unofficials support. The powers of the VC to deal with the violation of irrigation rules was handed over to the newly created Village Tribunals and later to the Rural Courts. Understandably, this was an attempt to introduce a British perception of the principles of justice to village affairs.

Today, all laws are subject to two important constitutional guarantees which ensure the rule of law and the fundamental rights of the individual. Judicial reforms have also resulted in impeding enforcement measures. At the same time it would be difficult at this juncture to bring back an arrangement by which representatives of farmers could be enabled to sit on judgement of matters which were originally included under the VCs. However the more feasible method appears to be to promote farmer organisations to bring social pressure on errant farmers as an extension of an conciliatory approach. As a last resort, action could be taken to fall back on legal procedures.

It is desirable to formulate legal procedures with regard to the need for rapid interventions and summary justice by officers who tend to take a more practical view of the problems encountered in the management of irrigation systems. This would mean that a court specially designated as a Water Court be set up to hold its sessions in the locality of irrigation systems on a regular visiting system. Court proceedings could be conducted without lawyers with the provision of appeal to a higher court.

Persons who preside over Water Courts will have to be senior officials in the Districts or someone selected from among the senior citizens who displays a proven capability to deal with these conflicts in an objective manner. It should be possible for these officers to be trained and appointed by the Judicial Service Commission.

Experience in Kimbulwana Oya in Kurunegala District indicates that social pressure can be effectively mobilized in bringing about a compromise. It is therefore necessary to ensure that conflict resolution be made an important function assigned to farmer organisations so that official interventions to initiate legal enforcement would be treated as a deterrent and as a last resort action.

Maintenance work

Proper maintenance of irrigation schemes by farmers was one of the principal considerations which motivated the colonial Government to revive ancient customs relating to paddy cultivation. But it always remained a vexed question. At the beginning, any improvement or repair to an irrigation system was subject to a recovery of the Government cost in 10 equal installments from beneficiaries and the imposition of an irrigation rate in perpetuity. The willingness of farmers to pay the irrigation rate was therefore made an important consideration in the administration procedure evolved for the purpose. The Irrigation Ordinance of 1935 relates the irrigation rate to both construction and maintenance. The method of recovery was administered initially by preparing a scheme for the operation and maintenance of the irrigation system. This scheme provided for the imposition of an irrigation rate, and for deciding on responsibilities between the Government and beneficiaries for maintenance, labour contribution, variation of rates, and conditions applicable to irrigation rates. it also provided for an exemption from rates in instances where beneficiaries agreed to undertake maintenance work on their own.

Subsequent amendments to the Irrigation ordinance show that the choice of the farmers to contribute by irrigate rates has been restricted by imposing the will of the bureaucracy. Apparently this resulted from a low collection rate because farmers were unable to honour the collective agreement with the Government. In fact the Director of Irrigation in the Administration Report for 1927 expresses his disappointment and reservations regarding the collection of irrigation rates.

More recently, irrigation rates or water taxes have been a politically sensitive areas of irrigation policy. A major revision of this policy was adopted in 1984 to enable the farmers to contribute towards the cost of operation and maintenance in the major irrigation schemes.

An important feature of this new policy is to ensure that contributions made by farmers will not be credited to a central fund nor allowed to finance any work outside the scheme. In effect, the new scheme attaches more importance to promote and mobilize farmer participation for maintenance than to the actual recovery of money in economic terms. In order to take this new scheme to its logical conclusion, farmer organisations are requested to identify maintenance items and to set priorities to prepare a maintenance programme for implementation in each year under the supervision of the ID. In the final analysis, the result would be to make the bureaucracy accountable to the farmer organisations and to the Project Committee for collection and disbursement of the O&M charges.

The subject of cost recovery cannot be easily cast in legal terms to suit implementation. As such, present experiences on the new policy will have to be monitored carefully to determine the best course of action. It is therefore necessary to set up broad guidelines in law for implementation with the maximum amount of flexibility for future adjustments.

Conclusion

In a social democracy, constitutional rights and guarantees are overwhelmingly important in safeguarding the rights of the individual. But these principles that help sustain an agricultural democracy do not necessarily apply with equal force to an irrigation democracy where a collective right to share a common resources is the primary concern. In conflicts associated with the equitable distribution of a common resources such as water, rapid interventions and decisions are of prime importance to safeguard the integrity of the system and the mechanisms which ensure equitable distribution.

To achieve these ends, it is necessary that irrigation legislation develop the maximum level of flexibility to accommodate such rapid interventions and decisions. Flexibility should be the hall-mark of irrigation legislation. These features have been recognized in many of the past Irrigation Ordinances. When attention was focussed increasingly on major irrigation systems which have different dimensions and magnitude to their problems, poor understanding of the complexities of such issues compelled the authorities to take resources to a path of least resistance by centralizing most activities in the hands of a bureaucracy and adopting highly uniform and rigid systems.

It is also true that no matter what policies and programmes are adopted, irrigation systems will have to keep moving, often due to the farmers who change and modify plans and schedules to suit their needs and perceptions. Even when wrong policies are adopted, the negative impact of such measures come to light long after the short-term gains have been achieved. Implementors of irrigation improvement plans get misled by these successes and repeat the same mistakes. Irrigation systems are often besieged by such short term policies. Even when such programmes are monitored closely, the true nature of their essential components has to be understood against a broad scenario of policies and programmes which link the past with the present.

Irrigation legislation by itself cannot bring about farmer participation. It can only spell out the broad framework for such participation and, to a limited degree, safeguard and facilitate the viability of the organisations in sustaining farmer participation.

Time has come to provide amendments to the current Irrigation Ordinance. Amendments Act No. 23 of 1973 was adopted to rectify certain legal impediments concerning repairs to damaged irrigation structures and jurisdiction of courts to try irrigation offenses. With the enactment of the Agricultural Productivity Law in 1973, the need to revise the Irrigation Ordinance to suit the new institutional order was highlighted over and over again but no action was taken. On looking back, this inaction cannot be regretted, although it may have happened for different reasons. It is contended that valuable information culled from filed experiences can provide the basic framework and the perspectives for a revision of the Irrigation Ordinance in the near future.

Appendix

A national policy for water must concern itself primarily with the optimum management of the country's renewable freshwater resources, represented by annual precipitation in the forum of snow and rain. Since most of this precipitation occurs during the relatively short monsoon season, it is necessary that we must aim at its conservation to the maximum possible extent so that we must aim at its conservation to the maximum possible extent so that water may be available for agricultural, domestic and industrial uses throughout the year. Since, further, the demand for water is not only already in excess of its supply but is also growing steadily in all sectors, we must make sure that this scarce resource is used with the utmost economy, and to the best possible advantage of the community.

Before there can be any meaningful discussion of how these twin objectives can be achieved it is necessary to consider how water behaves during the terrestrial portion of the perpetual hydrologic cycle. For we can make little use of water so long as water remains on or within the land mass that we can use it for our own purposes.

Of the precipitation which comes into contact with the surface of the land and does not get evaporated immediately, a portion runs off as surface flow while the rest percolates into the soil. The amount of water which runs off and the amount which passes into the soil and later into the sub-soil strata in the form of ground water depends upon a number of factors. Broadly speaking, the greater the intensity of rainfall, the greater the slope of the land and the greater its state of denudation, the greater will be the run-off and therefore the lesser the amount of water available to saturate the soil and recharge ground water aquifers. Further, since all flowing water dislodges and carries away some amount of the top soil, it also follows that the greater the run off the greater will be the erosive action of water.

It is extremely important that these basic facts about the interaction between precipitation and the land surface on which it falls should be properly appreciated because it is this interaction which holds the key to the understanding of almost all problems of water as well as of land management. For it is as difficult to manage silt-laden waters as it is to manage lands denuded of their top soil and therefore deprived of their rightful share of soil moisture and ground water.

Let us see why this should be so. The vicious circle of erosion has its beginnings in the destruction of the natural vegetative cover of the soil. Once this cover has been destroyed, whether by over grazing or by the conversion of forest lands into agricultural lands, and the soil laid bare to the fury of tropical rains, it begins to lose its fertility and therefore its capacity to support vegetation. For unlike water, the soil is for all purposes a non-replenishable resource—it takes Nature anything between 500 and 1000 years to build an icch of the fertile top soil. This is why erosion becomes more difficult to control with each passing year, and is accompanied by increasing aridity because denuded land surfaces offer little resistance to run-off.

The conservation of the soil is therefore necessary as much for its own sake as for the conservation of water. It is however equally necessary for the prevention of the damage which silt-laden waters cause in a variety of ways. For sediment not only eats into the storage capacity of expensive tanks and reservoirs but also makes the management of rivers an extremely difficult affair. For the carrying capacity of rivers gets reduced as sedimentation in their banks. Devastating floods and changes in the courses of rivers are thus the direct result of soil erosion. The construction of mebankments and dykes for the control of floods is not only an expensive business but offers no permanent remedy since they have to be continually raised in height to cope with the progressive rise in the level of river beds. The deposits of silt near the mouths of rivers choke bays and harbours and create acute problems of dredging and drainage in deltaic areas.

It is clear from what has been said above that a sensible water policy should concentrate on the conservation, to the maximum extent possible, of water as soil moisture an ground water rather than, as is happening now, on the storage of silt-laden waters in gigantic reservoirs and on the construction of expensive works for containing floods which are uncontrollable because they are bound to increase in fury with each passing year if their catchments continue to be neglected.

The rich dividends which a sensible water policy—sensible because it takes account of the realities of the relationship between land and water—and is not based on the purely engineering approach—would pay are easy to set out. Precious soil resources would be saved for the community-it has been estimated that the amount of soil which is displaced as a result of water erosion is of the order of 60,00 million tonnes annually and contains nutrients equivalent ot 5.37 million tonnes of NPK. floods, which cause losses of the order of Rs. 300 crores annually would be reduced in incidence and severity. The danger posed to tanks and reservoirs on which thousands of crores have been invested by sedimentation, would be removed. Finally, a great deal of water which is being lost to the sea would be conserved as ground water.

The benefits which additional supplies of ground water would confer on the community need special mention. For not only does this resource represents water in its purest natural form, but it also costs absolutely nothing to store for indefinite periods and what is more, to convey over the large distances which it travels under the surface of the ground in an effort as it were, to serve the largest possible number of people at their very door steps. Further, it suffers no losses by evaporation or seepage—either during storage or during transmission—such as surface waters are subject to. It is easy to tap as individual tubewells can be installed at little expense and within a matter of weeks if not in fact days. Once developed, ground water is available as and when required by the use on the mere pushing of a button and therefore lends itself beautifully to the requirements of modern agriculture.

By contrast, surface water is extremely difficult to store and manage. Big projects such as are the vogue in this country—take decades to design and build, require large areas of precious land for submergence and for distribution systems, are subject to serious evaporation and transmission losses, and demand stupendous outlays for their completion. They also create significant problems of water-logging and drainage and thus pose a threat to the very land they are meant to serve. Their storages are exposed to the threat of siltation and their command areas require large outlays for the full development of their potential. They require large staffs for maintenance and operation but find it difficult to give full satisfaction to the farmer in the matter of supplying water as and when required. Financially such projects are still in the red-the annual losses on their operation are of the order of Rs. 150 crores. But above all, they are wasteful of water—not only do transmission losses amount to around 40 per cent on the average, but since water is generally not charged for on a volumetric basis, the farmer is also extravagant in his use of this resource.

For all these reasons, every effort must be made to replenish ground water resources to the fullest extent possible. Apart from facilitating the natural recharge of ground water through soil and water conservation measures efforts must also be made to undertake artificial recharge wherever this is possible. Conditions must also be created to encourage ground water development to the fullest possible extent by carrying out land reforms and consolidation of holdings, by undertaking scientific investigations designed to ascertain the limits of safe pumping in given areas, by providing technical advice and guidance to the farmers, by arranging for loan assistance on an adequate scale and by making sure that tubewells do not remain idle on account of lack of electricity or diesel. In situations of limited supply of ground water, statutory powers must also be exercised to regulate development.

In an ideal situation, the planning of water conservation and use would be undertaken on a sub-catchments basis, in order to ensure that the water which flows into the main river system form each of its sub-catchments is not only free from silt but is also such as is genuinely surplus to local requirements. The calculation of local requirements would naturally depend upon the land use which is considered to be the best for the area in question and would involve a study of local soils, climatic conditions and available water resources. Arid areas should, in particular, not allow any run off to be lost and must utilize all precipitation locally—whether for growing crops or merely grasses and trees. Protection against over-grazing and over-felling must, in all sub-catchments, be undertaken along side such measures as contour ploughing, contour terracing and bunding and the construction of gully plugs and defenition weirs.

It is not difficult to imagine the dramatic change that the implementation of such an approach would bring about in the ecological environment that governs our fate as a nation. The denuded Shivaliks, the ravines along the Chambal, the Jamuna, the Sabarmati and Mahi would become things of the past, as would the treeless and grassless slopes of the Deccan plateau and the bare hills of North-Eastern India which have been stripped of vegetation by imprudent shifting cultivation. The millions of hectares of the so-called waste lands which yield nothing but silt and foods today would become important producers of grasses and would contribute significantly to the replenishment of the country's ground water resources. Our rivers would carry less sediment, and thanks to return flows from ground water aquifers, would maintain more even flows throughout the year. Our storages would acquire new leases of life and the severity of both floods and droughts would be reduced.

All this can however become possible only if the management of our water resources is not looked at as a problem of civil engineering but as something inseparable from the total management of our land and soil resources. If this is done, the conclusion will soon be reached that the first task before our Irrigation Departments today is to make the fullest possible use of the waters which have been already impounded at such heavy cost as this offers the best hope to the country for achieving self-sufficiency in food in the shortest possible time. This is however a task which requires a drastic reorientation in the attitudes of the irrigation engineer. Instead of dreaming of building more gigantic structures he must learn to pay attention to the detailed requirements of efficient irrigation and undertake the execution of the numberless small works which are required in each command to maximize production, and to prevent the wastage of water as well as damage to the soil. To illustrate, distribution systems need to be extended in a scientific manner over tens of millions of hectares so as to reach the last field in each outlet command, field need to be levelled and re-shaped in accordance with the contours of the land, and field as well as intermediate and major drainage provided wherever water-logging is or is likely to become a problem. A comprehensive approach to irrigation problems would also induce engineers to take more than a theoretical interest in the protection of reservoirs against siltation and would involve them seriously in soil and water conservation measures in catchments areas. All these are extremely complicated tasks and present a very big challenge indeed. They will also require huge outlays, running into thousands of course and will take decades to complete. The consideration of existing irrigation policies will thus not affect job opportunities for engineers in any way—if anything it will enhance them.

The new approach suggested above will also yield

some other benefits. By confronting Irrigation Departments with the total costs of projects—costs which include those of command area development and drainage as well as of reservoir protection—and by acquainting them with the difficulties in the way of planning and executing such "total" projects, it will bring them down to earth—both literally and metaphorically-and dampen their ardour for taking up big new projects even while existing projects are languishing. Such a change in attitude will be most welcome indeed and will hopefully cure Irrigation Departments of the disease of giganticism that most of them are suffering from. For today, Irrigation Departments think it below the dignity to even look at projects which cost less than Rs. 30 lakhs each—such projects are classified as "minor" and left to Panchayats and Agriculture Departments to plan and execute. Yet it is precisely such "minor" projects which are needed to be constructed in large numbers in the future if better use is to be made of our soil and water resources.

The new approach will also end the isolation in which the civil engineer has functioned so far in tackling irrigation problems—which by their very nature require the concerted attention of a number of disciplines. For it will no longer do to consider a project completed if the storage structure is completed and if the storage structure is completed and the distribution system built upto the outlets which serve blocks of land of anything upto 200 hectares each. If the heavy investments made in irrigation projects are to yield the results expected of them, the agronomist, the soil scientist, the extension worker, the drainage engineer and the economist will have to sit down with the irrigation engineer and consider what would be the best cropping patterns to adopt in various parts of the command, what would be the best way of meeting the water requirement of the crops to be grown, how excess water should be drained away and how various on-farm works should be financed.

In this background, the broad features of a water policy which might suit our circumstances would appear to be somewhat as follows.

(i) All land and water management problems must be viewed as a whole, as they do constitute an indivisible whole and cannot be considered in isolation from each other. Soil erosion, denudation, droughts and floods are all symptoms of the same discase of poor land and water management.

(ii) The natural unit for the planning and execution of all land and water management programmes is the sub-catchment which must retain for local use all the water which it needs. It is only if this principle is adhered to that it will be possible to build up rational River Basin Plans on the basis of plans for each of the sub-catch-ments within the basin. Needless, to say, such River Basin Plans will deal with land use matters as well as with the use of ground and surface water resources and the prevention and control or floods.

(iii) The conservation of water to the maximum possible extent must be attempted for obvious reasons. It must be realised however that the achievement of this aim is contingent upon the conservation of the soil. The prevention of erosion by water must therefore form a cornerstone of any water policy.

(iv) Since water is best and most conveniently conserved in the form of ground water, high priority must be given to the replenishment of ground water resources by natural as well as artificial means.

(v) The maximum possible development of ground water must be aimed at in view of the unique natural advantage offered by this resource. All the administrative, technical, financial and legal measures necessary for this purpose must be adopted as a matter of priority and the care of this resources should not be neglected any longer.

(vi) There is great scope for improvement in the working and utilization of the potential of existing and ongoing surface irrigation projects and therefore for a quick increase in our agricultural production. Priority attention must accordingly be given to the investigation of shortcomings in each project and thereafter to their rectification. Steps must also be taken to make each project financially viable.

(vii) New surface irrigation projects must be investigated with great care and steps taken to avoid the mistakes noticed in the planning and execution of similar projects in the past.

(viii) Land and soil are our most precious non-renewable resources and must on no account be allowed to be damaged by water in the form of floods or water-logging.

(ix) Water must be used with the utmost economy, whether for agricultural, industrial or domestic purpose. It must also be protected against pollution by agricultural, industrial or municipal wastes. Recycling techniques must be investigated and adopted wherever possible.

(x) The problems of water management are, by their very nature, not such as can be tackled by any single discipline. All the disciplines concerned with good land and water management should be brought together under new integrated Departments of Land and Water both at the Centre and in the States.

(xi) Water is a national asset and must accordingly be placed in the Concurrent List.

Growth of irrigation in India: An outline of performance and prospects

For the country as a whole, the proportion of net sown area under irrigation increased from 18.3 per cent in 1961-62 to 21.8 per cent in 1969-70. That is, the annual addition to irrigation was below 0.5 per cent of net sown area. The growth of irrigation from canals, which is predominantly a public source, has been much slower over this period. It increased from 7.7 per cent of the net sown area to 8.8 per cent. On the other hand, irrigation from wells, which are predominantly private, increased at a faster rate, i.e., from 5.4 per cent of the net sown area to 8.0 per cent.

The inter-State disparities in the percentage of net sown area irrigated is lower than that from individual sources indicating a significant compensatory influence of these individual sources. The inter-State variability or irrigation from wells increased significantly during this period whereas that from canals declined so that the variability in total irrigation remained the same. In *per capita* terms however the variation in total irrigation showed an increase.

In State like Gujarat, Maharashtra, Madhya Pradesh and Karnataka where the percentage of net sown area irrigated was significantly below the national average in 1961-62, the rate of increase was much higher than the national average. In Andhra Pradesh, Bihar, Tamil Nadu and West Bengal where the percentage of area irrigated was above the national average, the rate of increase has been smaller than the national average. However, Punjab, Haryana and Uttar Pradesh which were significantly above the national average in respect of irrigation, recorded much higher increases than the national average.

Prospects

For a tropical and sub-tropical country like India where agriculture is dependent mainly on erratic monsoons, and where uncertainty of rainfall diminishes with the rise in the level of rainfall, the availability of water through precipitation and artificial irrigation serves as a reasonable basis for a broad demarcation of agricultural development regions. States where bulk of the area accounts for low and medium rainfall and where the proportion of net sown area irrigated is significantly below the all-India average are classified as Dry Region. States with high rainfall are classified as Wet Region regardless of the proportion of sown area irrigated. States hiving medium and low rainfall but where the proportion of sown area irrigated is significantly higher than all-India average are also regarded as Wet areas.

Dry Region covers, by and large, Western India and Wet Region with high rainfall is located mainly in the Eastern Zone of the country. It is not an accident that both these regions which are essentially dependent on rainfall have remained relatively backward and Punjab, Haryana and Tamil Nadu accounting for low and medium rainfall but adequately served with assured irrigation have shown rapid progress.

According to the latest estimates, about 60 per cent of potential arable area can be ultimately irrigated from different sources and only about 42 per cent of this irrigation potential is likely to have been realised by the end of Fourth Plan. It will be of the remaining irrigation potential, major and medium projects account for as much as 65 per cent and ground water sources for only about 22 per cent. Ground water potential seems to be much less important in the States constituting Western Dry Region whereas it is relatively more important in the Wet Region. It will also be seen that the inter-State disparities in the ultimate irrigation potential are significantly lower when compared to the actual position in 1969-70. However, the inter-State variability in ground water potential is greater than that for major and medium projects.

Cost-benefit analyses of potential projects can alone reveal the extent of irrigation potential that can be profitably tapped. The area irrigated has been doubled in 23 years from 22.6 million hectares in 1950-51 to 44.9 million hectares in 1973-74. During the same period, agricultural output has almost doubled. By 2000 A.D., i.e., in an other two and a half decades, the demand for agricultural output can be expected to increase by nearly 100 per cent owing to the growth in population as well as the increase in *per capita* income. On the basis of past experience, if this requires the doubling of acreage under irrigation, another 45 million hectare will have to be brought under irrigation, bringing the total irrigated area to 90 million hectares or about 84 per cent of the potentially irrigable area.

The irrigated area may have to be increased at this rate despite technological change and increased use of fertilizers because the area under controlled irrigation which is conductive to modern intensive cultivation constitutes only about 10 per cent of the sown area at present, and it may be neither in the interest of efficiency nor of equality to promote a greater concentration of inputs in these limited pockets, which, in any case, may not be capable of meeting the increasing demand for agricultural output despite a rise in rices. Similarly, given a perspective of about two and half to there does not appear to be much scope for choice between different sources of irrigation or between projects located in different regions so long as they stand the test of social profitability.

Modernization of old irrigation projects is estimated to require an investment of about 900 crores which constitutes less than 10 per cent of the investment required for the new projects. Modernization of old projects may have a high pay-off because of the supplementary nature of such investments and because of the greater response of these areas to new technology. Paucity of investible resources with the government as well as considerations of social justice require that these prosperous areas are made to share a major portion of this investment, especially in view of their better resource position and because of high private benefits from modernization. However, considering past experience, the recovery of costs from them is unlikely owning to their influence both at the State and Central levels.

Eastern Gangetic Belt accounts for much of the untapped ground water potential. Its exploitation has been slow, especially on private account, partly because, owing to high rainfall in this region, irrigation is needed mainly for the second crop which might render such irrigation systems relatively costly and also because of the semifeudal social structure characterized by widespread poverty, low investible surplus, high interest rates and administrative inefficiency. All this is reflected, for instance, in the very slow growth of co-operative credit institutions in this region. Also, since rice is essentially a monsoon crop, in respect of which new technology has not made an appreciable impact, there may not be enough economic incentive, as yet, for a profitable exploitation of ground water potential. Therefore, public tubewells may continue to be the major source for the exploitation of ground water potential in this region.

Underground water potential as well as other minor irrigation schemes in the dry areas, especially in the western region where credit institutions are relatively well developed, are likely to be exploited with greater success in the next decade or so. However, such sources being costly, private investment in them can be undertaken only within a framework of high product prices as the subjective rates of discount would be high. Besides in the dry regions where minor irrigation potential is much scarcer than the potential from major and medium irrigation sources, large farmers may appropriate such water, despite the possible attempts to pre-empt such sources for small farms. As late comers, small farmers may have to incur higher cost for digging wells owing to the adverse externalities imposed, e.g., lowering water table, by the early exploiters. Indeed, in view of the possible adverse externalities for the early exploiters imposed by the late comers, there are already moves for imposing ceiling on the number of wells for each area.

Role of irrigation in integrated area development planning

Role of irrigation in agricultural development

The importance of irrigation as an essential input for agricultural development hardly needs any emphasis. Scientific practices such as the u se of high-yielding variety seeds, fertilizers, insecticides, etc., which have raised hopes for an ultimate solution of our chronic agricultural shortate, are all prmarily dependent on the availability of irrigation. According to one study, irrigation explains 54 per cent of the total variance in agricultural production for India as a whole. If Gujarat and Rajasthan are excluded from this estimate, the variance explained by irrigation goes up to about 70 per cent. Aside from making scientific practices feasible and thereby increasing per acre productivity of crops, irrigation also helps agricultural production by increasing the acreage under cultivation. Much of the fallow and cultural waste land, for example, can be brought under cultivation and multiple cropping can replace single cropping if irrigation is available.

Irrigation potential

The present crisis in agricultural production is one more dramatic illustration of the importance of irrigation in agricultural development. At the present time more than three-fourths of our net sown area are without irrigation facilities. Parts of the country which normally receive heavy rainfall and where the farmers do not consider irrigation as a permanent feature of their farming activities, are also not immune to sever drought conditions as has been illustrated during the last few years.

A quick glance at our irrigation statistics shows that

in 1951-52 only 18 per cent of our total cropped area was irrigated. The percentage went up very slightly to about 22 per cent and 25 per cent in 1969-70 and 1973-74 respectively. The estimated total potential for irrigation in the country is about 81 million hectares which is roughly 50 per cent of the total cropped area in the country. The Irrigation Commission has estimated that out of this total potential, 45.5 million hectares can be covered by Major and Medium Irrigation and 36 million hectares by Minor Irrigation Projects. These estimates are based on data supplied to the Commission by the various States and Union Territories and the are admittedly of a rough and ready nature. One can assume, however, that the estimated potential under major and medium projects is more precise as this is based on projected schemes in the public sector expected to be implemented in the near future.

As far as the potential for minor irrigation is concerned, the figure suggested by the Irrigation Commission may be an under-estimate. The reasons for this doubt are as follows:

(1) Not reliable data based on empirical evidence on the annual recharge and draft of ground water for the entire country are available, (2) estimates submitted by the States to the Irrigation Commission are based in known reserves only, (3) ground water can also be lifted in command areas of major and medium irrigation projects, (4) if resources are inadequate for constructing major and medium projects, a part or the whole of the area earmarked for these projects can be brought under minor irrigation.

Relevance of Major, Medium and Minor Irrigation in India

Aside from the fact that there may be an under-estimation

of the potential of minor irrigation in the country, there are other considerations which will make a strong case for emphasizing minor irrigation over major and medium irrigation in the Fifth and subsequent plans. Some of the more important considerations are discussed below.

Major and medium irrigation projects involve extremely heavy capital and operational expenditure. With present and, in all likelihood, future inflationary trends, it will be wise to review the projected allocation of funds for new major and medium projects in the Fifth Plan especially in the light of benefits that my be generated.

The time required by heavy projects before they become productive is extremely long. With the urgency of removing a critical agricultural shortage immediately and given the unpredictable weather conditions serious thought should be given to the appropriateness of committing scarce resources to heavy projects like these with long gestation periods.

Apart from the actual costs of construction and maintenance which are extremely heavy, all major and medium irrigation projects involve indirect and more serious costs to the society in relation to their benefits. The command areas of these projects almost invariably receive concentrated inputs for making the investment paying. It is well known that a large proportion of farmers in these command areas have adopted mechanized agriculture for fully utilizing irrigation facilities. This is understandable and is expected. However, areas falling outside the command areas remain backward. The economic disparity between irrigated areas with mechanized farming and the surrounding dry areas increases as the command areas become more prosperous.

Theoretically, intensive irrigation supported by mechanized farming in a few selected areas covering only about one-fourth of the total arable land of the country may eventually overcome the chronic food shortage of the entire nation. The philosophy behind the Intensive Agricultural District Programme for example, suggests concentrated efforts in a few carefully selected areas with assured irrigation. The results of the programme, in spite of the chronic shortage of essential inputs an bureaucracy intetia, have been impressive. In terms of quantity, one does not also doubt that the IADP districts have the potentiality for producing enough food for the entire nation. Yet at the same time, the economic condition of the vast majority of the agricultural population outside of the IADP districts, is becoming worse. In the absence of alternative non-farm employment opportunities the purchasing power of the non-IADP agricultural working force has reached such a low point that they are not able to buy food at current inflationary prices assuring food is available.

The alternative of the government procuring grains from surplus areas and recirculating them over a vast distribution network to the deficit areas has also not worked in the past. Even if by some miracle, the government is able to shoulder the burden, it will require extremely heavy subsidies to bring the price of the grains down to the level which is within the reach of the common man. No government can afford this, year after year, unless there is some hope of raising the purchasing power of the common man.

Selective intensive irrigation and programme such as IADP, although welcome for their favourable impact on good production, cannot by themselves solve the problem of scarcity which is a function of both the availability of food grains and the purchasing power of the people to buy them. Extensive porgrammes for the development of the dry and drought-prone areas must also be introduced simultaneously. This can only be possible if the available ground water reserves are fully tapped and extensive but decentralized minor irrigation projects are implemented.

The overwhelming majority of our farmers are small and marginal farmers. This is true in the command areas of heavy irrigation projects as well as in dry areas. While only a very few in the dry areas get the opportunity of irrigating their fields in the command areas, practically all farmers, small or big, have irrigation facilities. Yet, only the big farmers with sizable land holdings in the command areas and not the small farmers, get the full benefits or irrigation as the adoption of improved agricultural techniques is much higher among them. Several studies have discovered a high positive correlation between the size of farm and the adoption of improved techniques in the irrigated areas. The most important reasons suggested or this relationship are the higher risk-taking ability of the big farmers and their higher credit worthiness. It will be logical to conclude that the trend in these areas is towards larger land holding for minimizing the risks involved in mechanized agriculture and for increasing credit worthiness mainly based on the size of land holding. Since the amount of land is limited, larger land holdings of a few will either result in the reduction of the land holdings or even the displacement of others. Thus, an overall increase in agricultural production in areas with intensive agriculture, may benefit a few at the cost of others.

For many reasons, newly irrigated areas falling under the command of large irrigation projects do not automatically benefit the local farmers. Quite often, the new areas receive migrants form areas which have a long tradition of irrigation. The migrant slowly displace the local farmers due to their aggressive work ethic, resourcefulness, adaptability and skills. This phenomenon has been observed in practically all command areas of major and medium irrigation projects. The command areas of the Tungabhadra and Rajasthan Canal Projects are a case in point. In these areas, a large amount of land has also been transferred, legally or illegally to the migrants.

Heavy irrigation brings in other problems such as water-logging and salinity which also cost the society large investments in remedial measures.

Minor irrigation, on the other hand, costs less to the government as part of the expenditure is borne by the farmers themselves. Without doubt, loans are advanced by government agencies, commercial banks and other financial institutions but they are repayable with interest. The government is also spared a vast bureaucracy to run and maintain a large network of canals.

Trends based on available data would certainly indicate that he prospects of future agricultural development in India and an improvement in the general living conditions of our agricultural population depend heavily on the success of decentralized efforts which should cover prosperous as well as backward areas. Since our resources are already over strained, it will be impossible to continue heavy investments on major and medium irrigation projects and to launch a large programme of minor irrigation at the same time. Our position in this paper is that given the options and the cost-benefit differentials in a wider social sense, no new major project should be undertaken during the Fifth Plan unless one is already under construction. Barring exceptional schemes which are tied with other benefits such as flood control of generation of power or are located in hill areas where minor irrigation in the conventional sense of the term is not feasible, all proposals for new major projects should be postponed until the end of the Fifth Plan. In the meanwhile, a major proportion of funds thus saved, should be diverted to minor irrigation.

Importance of minor irrigation in agricultural development

In the light of our discussion of the suitability of major irrigation projects under Indian conditions, it will be easier to appreciate the important role minor irrigation can play in our future agricultural development. The primary advantage of minor irrigation is that irrigation facilities provided under such schemes can be decentralized as they are not contingent upon any single major source of surface water. Dry areas which remain almost totally dependent on the vagaries of the monsoon can be served by minor irrigation and each farmer can have his own source of irrigation water.

The cost of minor irrigation projects which is borne eventually by the farmers themselves, is also of manageable proportions. The credit required for constructing wells, tanks and installing pumpsets can also be distributed in a decentralized fashion.

The most important advantage of minor irrigation is that it can serve the small farmer directly. As the command area of an open well rarely goes beyond 5 acres, and average or small farmer can have assured irrigation for his land. This is in consonance with our accepted official policy of helping small and marginal farmers. The limitations inherent in the quantum of available ground water and the capacity of pumpsets would preclude the emergence of high landholdings if the cost of small farmers.

In the previous four Five Year Plans, allocations for minor irrigation have progressively been increased; in the Fist Plan, no provision for minor irrigation was made whereas in the Fourth Plan, minor irrigation was given about 30 per cent of the total allocation for irrigation. Percentagewise, although the allocation is still small, this certainly indicates the awareness on the part of the government of the importance of minor irrigation in our future agricultural development. The direct cost of minor irrigation schemes for the construction of wells, tanks, installation of pumpsets, construction of small field channels, as mentioned earlier, are borne by the farmer himself. Because of his inability to accumulate enough cash for these expenditures he is heavily dependent on credit facilities. Currently, credit for such purpose is being provided by the Agricultural Refinance Corporation, Commercial Banks, Land Mortgage Banks, Co-operative Societies and the various agencies of the State Government. Much of the credit which will thus be given to the farmer will have to come from government sources. This, however, will be repaid with interest in the future. The government in the long run does not, therefore, take the liability of the direct expenses involved in minor irrigation.

There are other expenses, however, which the government will have to bear and definite allocations for these will have to be made in the Fifth Plan.

These expenditures are:

(1) Cost for conducting scientific surveys of ground water resources in the country.

(2) Staff for periodic inspections for ensuring that the prescribed number of wells has been constructed and that the ground water reserves are not being depleted. This staff can, of course, be provided by the local Irrigation Department.

(3) Along with a survey of the ground water reserves, a thorough soil survey for the entire country, especially in the dry areas, should be undertaken. This will help in Agricultural Departments at the local level to advise the farmers in the most optimum utilization of the water resources by raising appropriate crops.

(4) Ground water will be available at various depths in the different parts of the country. Where water is at a great depth, deep tubewells will be necessary. This will be beyond the means of a single farmer and one could possibly think of Rural Water Co-operatives which could construct and maintain tubewells of this nature. As the command areas of such tubewells are bound to be quite large, the membership of the co-operative can also be large enough to sustain the co-operatives. Adequate financial help and assistance should be provided by the government to these co-operatives if necessary.

(5) There will be areas where ground water may not be available in economical quantities or the water may be saline and unfit for agriculture. In these situations, ground water lifted elsewhere could be piped to the deficient or saline areas. The cost of constructing such

pipelines will also have to be borne by the government.

(6) Technical problems involving large expenditures will be faced in hill areas where because of the topography, it may not be feasible to dig wells or bore tubewells. The numerous streams which are normally found in the hill areas may provide the source for minor irrigation in such cases and lift irrigation schemes on small-scale, out of reservoirs constructed at appropriate places can be thought of.

(7) One can conceivably think of strengthening the underground water bearing aquifers by channelising surplus surface water. This will require a through knowledge of the geo-morphology of areas deficient in ground water and the cost involved in channelising water from distant sources.

All of these will require funds and specific allotments for this purpose will have to be made. it is needless to emphasize that the first priority must go to a comprehensive scientific survey of ground water resources.

Minor Irrigation and Integrated Area Development Planning

The major problems with a large-scale programme of minor irrigation is however, neither financial nor technical as those can be solved. The problem will be in the proper utilization of the potential created.

The optimum utilization of the potential ground and surface water resources for the maximum impact on agricultural production will depend on the combination of a few essential factors. The availability of inputs such as credits, power fertilizers, pesticides and expert advice in right quantities, in right combinations, at the right time and at the right locations long with services for the post-harvest operations such as warehousing, marketing and processing facilities will ensure such utilization. Most of these-services and facilities are currently handled by individual departments and have to be co-ordinated in such a way that the farmer gets what he needs at the right time and at the right location.

The only viable basis for such inter-departmental co-ordination is to have integrated plans for small areas in which all departments involved in development work should participate. At the present time, each department has its own schematic budget which is spent on schemes formulated and executed without reference to the schemes of other departments. In additions, the pressure of spending funds within a prescribed time schedule results in duplications, wastage and a complete disregard of priorities. In integrated planning, one starts with the needs and potentialities of an area rather than with schematic budgets which have to be spent by a specific date in a year. If and inter-departmental effort is made to assess the needs and potentialities of an area as a whole, a comprehensive development plan for meeting the needs and developing the potential of the area will naturally follow. The various departmental schemes can then be formulated and executed within the framework of this plan and a well-co-ordinated time schedule can be followed.

The Planning Commission has recommended the preparation of integrated area development plans for each district in the country. A certain amount of work has already been done in regard to the development of appropriate methodologies for such planning. Actual planning for districts based on the principles of integrated area development has just started. This will be the time to reassess the future agricultural situation in different areas keeping the potential for minor irrigation in mind. Once the quantum of ground and surface water for minor irrigation projects is ascertained, one should be able to, after a close scrutiny of the soil types and other agronomic variables, project a cropping pattern which the farmers of a particular area, are likely to follow. Correctives to these projections will have to be introduced on the basis of the market situation and the demand for different crops.

The actual volume of future agriculture production can also be calculated on the basis of the probable availability of inputs such as credit, fertilizers, pesticides, etc. The projected volume of agricultural production thus obtained can be used as a basis for calculating the requirements for warehouses, markets and processing plants.

It is not the purpose of this paper to describe the actual planning techniques that can be used for integrating minor irrigation into agricultural planning or into area planning. We have underlined the need for doing so saying that the full impact of minor irrigation can be realised only if such integration is done. Minor irrigation, in order to be a viable proposition, will need a few inputs itself. A brief discussion of the more important ones is presented below.

Inputs Required for a Successful Minor Irrigation Programme

1. Credit

Currently, the Agricultural Refinance Corporation, the Commercial Banks, the Land Mortgage Banks, Co-operative Societies and a few other government agencies are advancing loans to the farmers independently or through agencies such as the SFDA, MFAL, GDA, etc.

With an enlarged programme on minor irrigation, a larger volume of funds will flow through these channels. A projection of the volume of funds that will be required in future will have to be assessed with reasonable accuracy. There are already too many agencies distributing credit to the farmers and no new agency is necessary. On the contrary, there is a great need for consolidating the work of some of the agencies so that the farmer can make application, get his loan sanctioned and take delivery of cash from the same agency. At the present time, the lack of co-ordination between various agencies and delay in processing applications, discourage many small farmers to make full use of the facilities available to them.

2. Power

Pumpsets for lifting water are run either by diesel or by electricity. Unfortunately, both are scarce these days and the farmer is unable to make up his mind about which energy to choose. Studies made on the economics of diesel operated pumps as compared to the electrical sets show a definite advantage of electricity over diesel. With the current petroleum crises and the escalating cost of diesel, the electric pumps should be more advantageous to the farmers. On the other hand, the current power shortage certainly provides a bottleneck for a large minor irrigation programmes. It is hoped that the supply of power will improve in the future as the solution of power shortage will have to be found internally. With a large reserve of coal deposits, many thermal stations can be constructed in the country. Diesel, on the other hand, will remain dependent on the international market for a long long time if not permanently.

The distribution of electricity to the farms also poses

problems. As lines have to be drawn to every village with potential ground water reserve, the State Electricity Boards find it uneconomical to pursue an extensive rural electrification programme. In order to assist the State Electricity Boards, the Rural Electrification Corporation provides loans for electrification with emphasis on the energization of pumpsets and small-scale industries. Loans of various kinds meant for ordinary areas, backward areas and for the minimum needs programme in selected areas, are advanced with staggered rates of interest. The principals and the interests are repayable over a long period of time covering up to 30 years. Although this timely assistance has eased the financial difficulties of the State Electricity Boards for the time being, the final repayments must be at least matched by the expected revenue. As is well known, agricultural power supply is heavily subsidised and has to be compensated by revenue from other sources such as industries. The question of raising electricity tariff, a sensitive political issue, is being debated in order to make rural electrification financially viable.

The question of financial viability of an input arises only when the revenue falls short of expected standards. The final yardstick, therefore, is increased agricultural production which will generate expected revenue. The whole issue rests not only on the optimum use of electricity and ground water but also on the successful co-ordination of all other programmes related to agricultural development. These programmes cover a whole range of activities including input distribution, storage, marketing and processing industries. The only way to ensure such co-ordination is to plan for it and implement it successfully. Integrated area development plans, if properly formulated and implemented, will provide a method by which the various inputs which contribute directly or indirectly, can be made to pay for themselves as they help to increase the overall agricultural production in the country.

Integrated approach to utilisation of surface and underground water

The basis problem of developing countries in relation to water resources development is essential to ensure the best possible economic and social utilization and this process of optimization must be carried out within the financial constraints imposed by the scarity of capital. The ultimate insistence is on the maximization of benefits while minimizing expenditure. But the long-term perspective in the planning of the water resources development and utilization must not be lost sight of and a balance has to be struck in the utilization of surface and ground water resources. An inter-disciplinary approach to the matter must be evolved so that a surface water specialist may not show unawareness of the complementary role that could be made by ground water. There are certain regions/river basins where bulk of surface water resources have been tapped for utilization and in such cases the ground water potential has to be created not only to further increase the irrigation potential but also to augment and sustain the created irrigation potential of surface water resources, which remains susceptible to erratic monsoon rains.

The development and use of underground water resources on a comprehensive scale has always been a particularly difficult proposition for planners, primarily because it is a source that has to be fully explored and studied in considerable detail. However according to the latest available report, the utilizable surface run off and utilizable ground water potential are respectively 666 thousand million cubic meters. The utilized quantitites are 240 thousand million cubic meters and 204 thousands million cubic meters and 96 thousand million cubic meters through gravity and lift irrigation schemes. In contrast to surface water resources, the ground water is usually not restricted to sharply defined channels or basins, but is present almost everywhere in the saturated zones of subsurface geologic formation. In many regions, it constitutes a single water reservoir that may extend beneath a number of surface water basins that are truly isolated from each other. The ground water accounts for more than 95 per cent of all the liquid freshwater available on the earth at any given moment. It can be found in the uppermost layers or rocks and sediments almost everywhere, even in many desert regions.

The pre-requisite to formulating a national policy on utilization of surface and underground water is to carry out a comprehensive study of the water potential, which should cover the following items.

(i) Assessment of the seasonal and annual run off of the various river basins and their sub-basins and their dependence on climatic fluctuations and precipitations.

(ii) Comprehensive study of ground water reserves; study and classification of the hydrogeological model of the various aquifers, their sources of replenishment and modes of drainage, constraint to their utilization, such as sea water intrusion, salinization, depletion, etc:

(iii) Interaction between surface flows and the ground water aquifers.

The changes and fluctuations occurring in natural phenomena like precipitation, temperature, etc., add to the inaccuracy of hydrological findings and conclusions. Statistical methods have to be taken resort to for obtaining reasonably dependable mean values, but this also calls for the collection of relevant date covering good number of years. But preparation of a water resources inventory is indispensable before evolving an integrated approach to utilization of surface and ground water for irrigation.

The utilization of ground water undoubtedly involves fewer losses and the efficiency of its exploitation is usually much higher than that of the surface flows. This is because the water bearing strata of the ground represent a natural storage facility of wide dimensions and permit exploitation in exact accordance with the actual demand at any time. In contrast to this, the canal command areas in most parts of the world are still being irrigated by the same old wasteful methods, i.e., the flooding, furrow or basin irrigation methods. The experience of age-old systems of irrigation in India as in other countries has prominently shown that irrigation is not an un-mixed blessing. The history of irrigated agriculture in India shows that uncontrolled and overall irrigation makes for wastages of water resulting in as low an irrigation efficiency as 25 to 30 per cent in the case of major canal projects and 30-35 per cent in the case of medium canal projects. It needs hardly to be over-emphasized that such a practice not only leads to wasteful use of water but also causes rise in the water table of the area eventually leading to water-logging and salt efflorescence to the foreground. To forestall such a dangerous situation, it would be most appropriate to adopt a conjunctive use of surface and ground water or optimum irrigation benefits. The simultaneous introduction of tubewell irrigation in such canal ayacuts would progressively check the rise of water table, as also will be augmenting the canal irrigation during unfavourable periods of surface water flows. Where co-existent, the exploitation of surface and ground water has to go together in a well planned, phased and co-ordinated manner. Failure to do so will not only lead to duplication waste of natural resources but also upset the natural equilibrium of surface and sub-surface flows. In fact, a judicious combination of canal and tubewell irrigation system offers an ideal solution to the problem of water-logging which may be apprehended in course of time with the introduction of perennial irrigation in a major canal project.

There is one more important factor which plays an important part in checking the rise of water table in a canal command and that is controlling "intensity of irrigation" or permitting only a fraction of the total culturable commanded area to receive water or limiting annual irrigation area. Areas with high water table may be allowed to receive kharif irrigation only, when the water table may rise temporarily, but during rabi, canal water irrigation may be drastically cut down so as to deplete the water table. However, irrigation facilities can be extended to rabi crops through tubewells, which would effectively lower down the water table also. It is in this manner that a coordinated and phased approach to the utilization of the surface water and underground water resources can be evolved.

Lift irrigation is the only alternative for extending irrigation benefit to such scattered high patches in the command of a canal project which are not covered by gravity flow. The exploitation of ground water resources for such areas would also serve as an anti-water logging measure of the canal command.

To sum up briefly, it can be said that a comprehensive assessment of the available surface water and ground water resources has to be made basin/region-wise as also the existing utilization of the same. For the basins/regions where the surface water resources have been mostly tapped, it is but expedient to exploit the ground water resources extending irrigation benefits to the new areas and augmenting the existing ones. However, where there is enough surface water resources remaining to be exploited, the simultaneous tapping of ground water resources in a planned, phased and co-ordinated manner is nevertheless essential. An integrated approach to the utilization of surface and underground water is a wise proposition for getting optimum benefits of irrigation on a long-term basis. However, there are certain limitations in the exploitation of ground water resources, e.g., shortage of power in certain regions which may stand in the way of launching ambibious programme of lift irrigation. Moreover, since power is a rare commodity, being so much in demand for major industries and small scale industries that may be in the offing, it is a big question mark whether the same could be diverted for exploitation of ground water resources for such areas where surface water resources are in plenty to be tapped. Where of course flow irrigation is not feasible at economical cost, the utilization of ground water is a must and this would generally have a over-riding consideration for use of power.

In view of tight financial resources, the economics of the surface and underground water development schemes would also have to be given due consideration. It has to be admitted that though the capital cost of a canal scheme may be more in certain cases, in view of the much lower maintenance cost and also much longer life, the canal irrigation scheme may be a more viable economic proposition.

Index

Irrigation management, farmer participation in, 233